The Machine Penalty

"The age of AI is here, but a striking 'machine penalty' remains—people's deep-seated reluctance to trust AI with decisions about morality, relationships, and beyond. What drives this distrust, and will it ever fade? With a careful synthesis of cutting-edge research and vivid examples, Daniel Shank leads readers through a fascinating investigation into the barriers between humans and machines, revealing why it matters for our lives, and what's at stake for our future."

—Kurt Gray, *Professor of Psychology and Neuroscience at the University of North Carolina at Chapel Hill, Author of Outraged: Why We Fight About Morality and Politics and How to Find Common Ground and The Mind Club: Who Thinks, What Feels, and Why It Matters*

"Shank's nuanced account of human-AI interaction is at once tightly scoped and expansively relevant, carving a new research agenda that advances AI studies, social psychology, and especially, their intersection."

—Jenny L. Davis, *Professor of Sociology and Gertrude Conaway Vanderbilt Chair at Vanderbilt University, Author of How Artifacts Afford: The Power and Politics of Everyday Things*

"The Machine Penalty presents a timely look at human interaction with technology and is written with considerable rigor from an expert in the field. An important read for those who seek to make sense of our increasingly automated age."

—Adam Waytz, *Professor and Morris and Alice Kaplan Chair in Ethics and Decision Management at Northwestern University, Author of The Power of Human: How Our Shared Humanity Can Help Us Create a Better World*

"Daniel Shank adeptly synthesizes from a broad range of social psychological and human-computer interaction research to take on one of the newest, most important conundrums in our world: how we perceive and interact with artificial intelligences. This is a must-read to understand the dawning age of human-AI integration."

—Janet Pauketat, *Research Fellow at the Sentience Institute*

"I was fascinated as I read *The Machine Penalty: The Consequences of Seeing Artificial Intelligence as Less Than Human*. It reveals the hidden biases that lead us to undervalue AI outcomes, even as technology matches or surpasses human capabilities. Daniel Shank, an expert on the psychology of technology, explores how these tendencies impact everything from fairness to creativity, and influence our society in ways we may not realize. This compelling book challenges us to rethink our interactions with AI and the profound consequences of treating machines as inherently 'less.'"

—Nathanael Fast, *Associate Professor at the University of Southern California and Co-director of the Psychology of Technology Institute*

"Shank masterfully integrates the growing research on how humans perceive machines, and societal and psychological implications of these beliefs. Engaging and thought-provoking, this book is a must-read for anyone curious about the relationship between humanity and technology."

—Yochanan Bigman, *Senior Lecturer at the Hebrew University Business School*

"*The Machine Penalty* not only effectively reviews the most current scholarship on how we perceive and work with technology; it distills these research trends into actionable insights. Anyone interested in the psychology of technological change will appreciate Dr. Shank's persuasive summaries and examples."

—Arthur Jago, *Assistant Professor of Management at University of Washington Tacoma*

Daniel B. Shank

The Machine Penalty

The Consequences of Seeing Artificial Intelligence as Less Than Human

palgrave
macmillan

Daniel B. Shank
Psychological Science
Missouri University of Science and Technology
Rolla, MO, USA

ISBN 978-3-031-86060-7 ISBN 978-3-031-86061-4 (eBook)
https://doi.org/10.1007/978-3-031-86061-4

This Palgrave Macmillan imprint is published by the registered company Springer Nature Switzerland AG.
The registered company address is: Gewerbestrasse 11, 6330 Cham, Switzerland

To Mom, Dad, and Brother,
Writers, Readers, and Scrabble Players

Competing Interests

The author has no competing interests to declare that are relevant to the content of this manuscript.

Contents

About the Author

Daniel B. Shank is Associate Professor of Psychological Science at Missouri University of Science and Technology, specializing in the areas of social psychology and technology. His research primarily focuses on social psychological interactions with and perceptions of artificial intelligence, including morality, emotions, relationships, impressions, and behavior toward AI.

He has received over $3 million in external and internal grants, including major external grants from the *National Science Foundation* and the *Army Research Office*. He has published over 50 peer-reviewed articles in psychology, sociology, and behavioral science technology journals such as the *Computers in Human Behavior; Journal of Personality and Social Psychology; Journal of Experimental Psychology Applied; Technology, Mind, and Behavior; Social Science Research; American Behavioral Scientist*; and *Human Factors*. He's received awards for Research, Teaching, Student Impact, Experiential Learning, and overall Excellence from Missouri S&T.

Dr. Shank previously received a BA in Computer Science from Harding University, an MS in Artificial Intelligence from the University of Georgia, and an MA and PhD in Sociology from the University of Georgia. Before joining Missouri S&T, he was a postdoctoral research fellow at the University of Alabama in Birmingham and at the University of Melbourne in Australia.

List of Figures

List of Tables

1

Introduction

Imagine you've just received a really positive recommendation for a restaurant. Does it matter who makes the recommendation? You might assume a critic's recommendation is more valuable than a random customer's. Your close friend might know your tastes and give you better advice, compared to a stranger. But what if the only difference was that the recommender is either an artificial intelligence entity (AI) or a human? All else being equal, you may consider that an AI doesn't really understand food, nor the experience of eating out, leading you to prefer the recommendation by your fellow human.

Now consider this: you've just heard about a loan company that has been systematically denying loans to qualified individuals of some races but not others. You'd probably think this is completely unfair and may look for someone to blame. Would you find this outcome any fairer if you found out it was caused by a human manager compared to an AI-based loan program? Likely you would be fine blaming the human manager as incompetent at best or racist at worst. However, if you're like most people, you wouldn't blame the AI as much—after all, it's just making its best decisions based on its data and algorithm. You might even think that the outcome wasn't all that unfair since it was decided by a program.

D. B. Shank, *The Machine Penalty*, https://doi.org/10.1007/978-3-031-86061-4_1

For restaurant recommendations, denying loans based on race, and a number of other situations, we tend to see outcomes from artificial intelligence as less than those from humans. I call this the machine penalty.

The core of my argument, simply put, is this: *comparing AI to humans leads us to diminish similar outcomes from AI across situations.* This idea might seem odd, as AI and their outcomes often seem *greater than* humans—faster, superior, more accurate—in areas as diverse as writing, evaluating, summarizing, producing, gaming, calculating, and creating. And aren't they only getting better every year, if not every month? How then can AI be *less than* humans?

The key is that the machine penalty is not an argument about AI's actual abilities or capacities; it's an argument about comparisons and perceptions. It's all about how we view AI and what we believe and infer about it. In spite of capacities and developments in artificial intelligence, it's an argument rooted in the fact that people know that AI is still not human, and as such, we as humans tend to—rightly or wrongly—compare it to ourselves and assess their deficiencies and differences through that comparison. In the restaurant recommendation example, this comparison stems from AI having less food experiences based on not having a biological body. Whereas in the loan example, it's related to having less ability for prejudice based on making decisions on computational and objective criteria. Regardless of the specific example, it's clear that we perceive humans and AI differently, and throughout this text, I explore how that works and what its implications are in our increasingly AI-driven world.

1.1 The Machine Penalty: AI < Human

The penalty is the argument that *comparing AI to humans leads us to diminish similar outcomes from AI across situations.* Notice that the diminishing is not a blanket comparison between all AI and human outcomes, but, instead, critically applies to situations with *similar outcomes.* There are many tasks wherein the outcomes are not at all similar, and the penalty wouldn't apply to them.

An illustration of dissimilar outcomes would be comparing a human to a wheeled robot: if the outcome is getting up the stairs, the human wins; if it's a race on a track, the robot wins. In this case, there is no penalty as the outcomes are not at all similar. Either the human or robot is superior: end of story. Likewise, a company's chatbot might win on speed for answering questions related to the company's services, whereas a company's human customer service representative might win on breadth of advice, being able to suggest services well beyond the company's core options. In these and many other cases, the outcomes are quite different, and the machine penalty doesn't apply. However, in other situations, an AI and a human might have similar capacities, and, in many cases, their outcomes are the same—or extremely similar—and then the penalty would apply.

These examples bring up two additional points about the penalty. First, the outcomes one cares about are dependent on one's objectives in a situation, such as whether the goal of talking to a human customer service agent or chatbot is speed, breadth of advice, or information. Second, the penalty is less likely to apply to the more objective outcomes of speed, price, quantity, accuracy, correctness, success, and winning, as slight differences in these outcomes make them dissimilar. In contrast, the penalty is more likely to be applied to the more subjective outcomes of trust, quality, liking, influence, accuracy, genuineness, appropriateness, authenticity, personalization, fairness, morality, goodness, and power. For these, it's quite possible to have similar outcomes, if not identical. Additionally, these subjective outcomes can more easily shift in value—that is, be diminished—because they are more subjective. In Chap. 2, I expand on these arguments by more fully theorizing about the penalty, including detailed conditions of when it does or does not apply.

In this book, I situate the machine penalty within a larger argument that links the penalty to its causes and consequences. The root causes of the machine penalty involve people comparing AI to humans and perceiving differences and often deficiencies. AI are like a different species that never fully reaches the benchmark of *humanity*, even though in specific capacities, AI have exceeded humans. I characterize these comparative differences as the root causes of the penalty because, across situations

Fig. 1.1 The machine penalty is caused by human–AI comparisons and specific interaction situations and leads to immediate and downstream consequences

and domains, we must first compare humans and AI in order to diminish them. That is, if we don't perceive AI any differently than humans, there would be no cause to diminish their outcomes. Aspects of specific situations also cause—and interact with these comparisons to cause—the machine penalty. Some situations are highly likely to produce a penalty, whereas others aren't at all or might even reverse it.

The penalty produces consequences, both immediately in those situations and downstream implications for society. I'll now briefly expound on this basic argument, visualized in Fig. 1.1, which will serve as an outline for the rest of the book.

1.2 Human–AI Comparisons as Causes

The root cause of the machine penalty stems from a comparative assessment where we perceive AI are not quite human. AI are essentially a different species and, compared to typical human benchmarks, we find they are not identical or equal. Even when they are similar in many ways—but especially when they display inferiority, superiority, or just dissimilarities in specific domains—they often miss this human benchmark, thereby appearing distinctly nonhuman. Essentially, our comparison answers the question: *How do we perceive AI as distinct from humans?* As in earlier examples, these differences might include our assumptions that AI do not have human experiences, a biological body, or inherent prejudicial bias, while also being more systematic and objective. This question, however, suggests an even more rudimentary question: *How do we perceive machines?* I take up this last question in Chap. 3 by introducing the major social psychological theories that address how we perceive machines.

Then, in Chap. 4, I return to answer the former question, suggesting five primary human–AI comparisons: appearance, identity, behavior, mind, and essence. First and most obvious, AI, robots, and machines have different bodies, voices, faces, and movement than humans and from each other. Second, AI identities, along with names and labels, are assigned to them often based on what they are designed to do, whereas humans are the ultimate generalists, being able to take on different social and personal identities. Third, AI behaviors are often a key difference from humans, including different capacities based on their bodies—or lack of bodies—as well as quicker but more formulaic responses. Fourth, AI have different minds than humans, including a different type of thinking and much less capacity for feeling and experiencing the world. Fifth, people perceive AI having a different essence or fundamental nature than humans, such as being mechanical, incapable of true emotion, and not being sentient.

These human–AI comparisons are considered the root causes of the machine penalty because if we don't believe there are any differences between AI and humans, we have no basis upon which to diminish their outcomes. Consider this analogy: you are evaluating two rudimentary drawings of cats, both recognizable but not wonderfully accurate, and, overall, fairly similar in quality. With no other information, we might consider them as equals. But we might think the three-year-old who drew one actually displayed more artistic skill than his seven-year-old sister. In this case, this age-based comparison is the root cause of different judgments—diminished or enhanced judgments—because they set up different expectations. We expect three-year-olds to have different, less developed fine motor skills compared to seven-year-olds, and this comparison is at the root of our evaluation. Likewise, the human–AI comparison is the root cause—but not the only cause—of the machine penalty.

1.3 Interaction Situations as Causes

The human–AI comparison answers the question, *How do we perceive AI as distinct from humans?* Holistically, all the causes of the machine penalty address the question, *What leads us to diminish machine outcomes?* I begin answering this question by looking at the evidence that we do, in fact, often diminish machines and when, how, and why we do that. In Chap. 5, I address this with a broad lens by looking at examples from history, my own early research, and the current research on *algorithm aversion*. Then, in Chap. 6, I overview the research evidence on what causes the machine penalty. This includes five human–AI comparisons as causes but also shows five aspects of the situation that are critical to causing—or not causing—the penalty.

These situational causes include whether the situation involves controllable, personal, important, subjective, or moral decisions. The more control we desire in a situation, the more we diminish AI's outcomes compared to other people's, especially our own. If the situation involves personally important or personalized outcomes, we also prefer humans to machines. When the situation is less important or mundane, we diminish AI's outcomes compared to human's. Situations or the outcomes of them that are more subjective or taste-based enhance the machine penalty, compared to more objective, quantitative outcomes, which we perceive as appropriate for AI. Finally, situations that deal with morality and human life itself seem less appropriate for an AI and more appropriate for a human, thereby increasing the likelihood of a machine penalty.

1.4 Applying the Penalty and Its Immediate Situational Consequences

As already detailed, Part I (Chaps. 2, 3, 4, 5, and 6) of this book provides the background, research evidence, and causes that lead to the machine penalty at large. Then, in Part II (Chaps. 7, 8, 9, 10, and 11), I proceed to apply the penalty to the real world in five different domains, along with its immediate consequences within those domains.

Because AI functions in many roles traditionally filled by humans, the machine penalty applies to AI and human outcomes across a number of different areas and situations. While the basic principle is the same across domains, the particular outcomes that are diminished, the conditions in which the machine penalty is most likely to apply, and the consequences vary considerably by domain. Specifically, I demonstrate that we penalize machine outcomes when they give advice (Chap. 7), evaluate people (Chap. 8), cause harm (Chap. 9), produce art (Chap. 10), and provide companionship (Chap. 11).

Supported by research evidence, I show the machine penalty at work in each of these applied domains for critical outcomes related to that domain. Yet in some cases, the research shows that specific conditions lead to no penalty, or even reverse it, creating a machine advantage. For those cases, I show how the ten causes—both comparative and situational causes—explain the initially conflicting evidence. In Part II, by tracing those domain-specific conditions, the research shows which AI outcomes are diminished, when that occurs, and the immediate implications within that domain.

Some of this domain specificity is because there is a primary outcome that is more central to that domain and thus most critical for the machine penalty. For AI giving advice, the primary outcome is influence, where machines exert less influence on subjective and personal advice, and we are less likely to choose an AI to advise us in those areas. When AI evaluates people, the primary outcome is fairness, where AI decisions are seen as less fair than humans evaluating other people for hiring, firing, promotion, criminal sentencing, and policing in general. Yet if the outcome is known to be biased—as in the opening example—we then find the same biased outcomes as less *unfair* when coming from an AI.

Although AI and robots are typically programmed to mitigate harm, when they do cause harm, blame is the primary outcome. We typically blame AI less than humans for the harm they cause. When AI produces art, the primary outcome is the artwork's value—both monetary and aesthetic value—which are often penalized for art created by AI based on their identity and authenticity. For AI that provide companionship, the primary outcome is satisfaction with the relationship, where machines are usually perceived as less satisfying in their conversations, interactions, and authentic companionship.

In a theoretical sense, the penalty represents an error in judgment. Therefore, succumbing to that error is a human bias, whereas correcting for it provides us the most accurate view of the world. Yet, in a practical sense, the machine penalty's immediate consequences can be either good or bad, depending on whether a positive or negative outcome is penalized. Diminishing AI's desirable outcomes is bad, while diminishing their undesirable outcomes is good. Specifically, an influence penalty for advice can reduce or enhance performance, a fairness penalty for evaluating people can check or fail to check bias, a blame penalty for harm can erode or increase safety, a value penalty for art can cheapen or elevate creations, and a satisfaction penalty for companionship can lead to real or fabricated love.

1.5 Downstream Societal Consequences

The domains lead first to those immediate consequences regarding their primary outcomes: diminished influence, fairness, blame, value, and satisfaction. But beyond these individual instantiations where a machine is penalized, what are the broader, more diffuse effects of treating machines differently than humans? In Part III (Chaps. 12, 13, and 14), I resolve the greater influence and impact of the penalty by considering its downstream societal consequences. These consequences reach across human institutions, including implications for people, businesses, truth, uniqueness, and relationships, which is explored in Chap. 12.

These consequences suggest both warnings and opportunities for society including ethical and logistical issues stemming from our perceptions of AI. I suggest how we might respond to the penalty and its consequences in Chap. 13, in which I lay out general recommendations based on living in a world where we interact with and tend to diminish AI's outcomes.

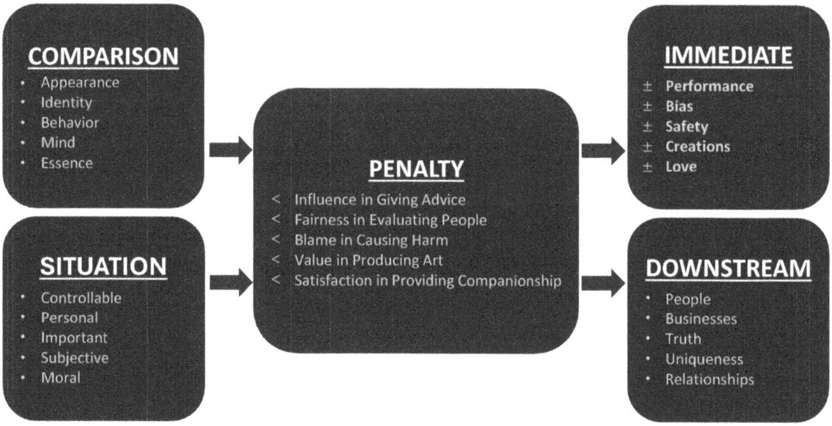

Fig. 1.2 The full machine penalty argument, including human–AI comparisons and situations as causes, the penalty, and the immediate and downstream consequences. The box for the penalty lists outcomes that are less for AI compared to humans in each domain, while the immediate consequences box lists potentially positive or negative consequences based on applying the penalty in each domain

Chapter 14 concludes by considering future AI and how the machine penalty might be applied to them. Combining what I have overviewed so far, Fig. 1.2 visually depicts the elements of my entire machine penalty argument.

Part I

The Penalty Explained

2

What Is the Machine Penalty?

2.1 Defining the Penalty

Let's unpack the statement, *comparing AI to humans leads us to diminish similar outcomes from AI across situations,* by considering the meaning of *AI, AI versus humans, across situations, outcomes,* and *similar outcomes.* Expounding on each one these terms and phrases in detail will help lay out the conditions of the machine penalty—the broad range of situations to which it applies—and clarify definitions for terms I use. Formal condition statements show what situations are outside the penalty's scope and, therefore, may be best explained in other ways.

2.1.1 What "Artificial Intelligence" Includes

There are many definitions of artificial intelligence, machine learning, and artificial agents, and even some disputes about what counts as a robot. Additionally, artificial intelligence isn't a static concept in terms of the techniques it includes, as there have been shifts in techniques from symbolic AI, expert systems, and genetic algorithms to deep-learning neural networks, machine learning, and generative adversarial networks.

D. B. Shank, *The Machine Penalty*, https://doi.org/10.1007/978-3-031-86061-4_2

These have given rise to the popular large language models and generative AI techniques, that have recently become well-known even outside AI research circles.

However, those aren't central to the machine penalty argument. We often don't know—and don't have to know—what's "under the hood" of an AI system to perceive and interact with it. As we'll see in Chap. 3, we can perceive any sort of technology, even non-digital technology or even non-technological objects in meaningful ways that cause us to judge it. The reason I focus on AI instead of, say, just computers, is that many of these human–machine comparisons are now only possible because AI can do similar tasks as humans. Some of the slightly older work that I build on does refer to computers, computer systems, or computer programs, because they used simple programs and "AI" wasn't a household term at the time.

Physical embodiments including robots and other machines, digital representations including avatars and characters, as well as completely disembodied AI systems including AI programs and algorithms are all subject to this machine penalty argument. The differences among those are still critical, as they form the basis for human–AI comparisons covered in Chap. 4, which are root causes of the machine penalty covered in Chap. 6. Overall, I'll use AI, machines, robots, or similar terms throughout this book. I'll also usually refer to AI as "it" or "they," but I will use gendered pronouns when discussing AI that have been gendered, either in their design or people's perceptions of them. Similarly, I'll usually use "AI," to refer to all artificial intelligence, multiple AI entities, or specific machines.

But most critically, for some perceiver to apply the penalty toward a machine, the person must perceive AI as its own entity or decision-maker, even if it's not technologically sophisticated. An example of how this applies is Microsoft Bard, which is built into their search engine but often shows an animation of the text being composed. This feature seems to have the purpose of helping us see and anthropomorphize Bard as its own entity instead of just another search result, even though searches themselves are a sophisticated technology. Therefore,

Technology Entity Condition: The penalty applies to any AI, machine, program, or other technology that is perceived as its own entity which can cause an outcome.

2.1.2 AI Versus Humans

Next, the outcomes that are diminished come from *AI* instead of *humans*, and this delineation can apply in five specific ways.

First, the penalty does not apply if you don't know whether an outcome has come from an AI or a human.

Second, it also would not directly apply if a human and AI are jointly producing an outcome. Hybrid human–AI teamwork is important, and is covered more in the Chap. 14 conclusion, but the machine penalty wouldn't directly apply because the outcomes are jointly produced, not separate and comparable.

Third, this would apply—albeit incorrectly—in situations where a human is passing for an AI or an AI is passing for a human, such that the penalty is about who or what we believe caused an outcome, not who or what actually caused it. Experiments often manipulate these labels to separate the perceived agent from the actual outcome.

Fourth, in specific situations, we may compare one AI or group of AI agents to one human or a group of humans. The comparisons could also be more or less proximate or specific in time or entities selected for comparison. For example, a new AI system created to do improvisational comedy could be compared to previous human improv shows in general, to a specific group of current comedians, or it could go head-to-head with Colin Mochrie of *Whose Line Is It Anyway?*

That leads to the fifth, which is by selecting different human(s) or AI(s) we change the application of several other conditions. Specifically, this section changes the situation (*Situational Effect Condition*), thereby changing the outcomes (*Relevant Outcomes Condition*) and potentially the outcomes' similarity (*Similar Outcomes Condition*). Taken together,

AI vs Human Condition: The penalty applies when one believes some outcomes are caused by AI(s) versus human(s).

2.1.3 Across Situations

As a broad theory that can be applied to different types of AI versus human comparisons, the machine penalty isn't present in every instance. *Across situations* implies that it is a general principle, but there are cases that change its applicability. But we also have a good idea on situations where it is more likely and those where it is less likely, or even completely reversed. The situational causes in Chap. 6 provide some clear evidence that the situation itself causes or alters the penalty based on specific characteristics of the situation.

Consequently, we might broadly consider that the penalty *could* occur in any situation that meets the criteria laid out in these five conditions. In other words, lots of situations are within the scope conditions of this theory. Yet, we expect some of those situations to be ones where the penalty is strengthened and highly likely to occur, and others where it is weakened and less likely to occur. Some are even expected to reverse it into a machine advantage. Likewise, the specific situation constrains what outcomes are applicable, as we'll see in the next two conditions. Thus,

> *Situational Effect Condition: The penalty applies across a range of situations, but some types of situations are expected to strengthen, weaken, or even reverse the penalty.*

2.1.4 Outcomes

Outputs are the decisions, products, communications, information, interactions, or other behavior directly caused by an AI or a human, in line with the *Technology Entity* and *AI vs Human Conditions*. This is differentiated from *outcomes*, which are the aspects of those *outputs* that are important and relevant to a situation, in line with the *Situational Effect Condition*. For example, the output of an improvisational comedian AI is the comedy routine, whereas the outcomes could include how funny, how clever, or even how long the routine is. These outcomes are what specifically is diminished. They could include trust, quality, liking, genuineness, appropriateness, authenticity, personalization, fairness, morality, goodness, or potency.

It's also possible and likely to have multiple types of outcomes for the same situation. If an AI or human hiring manager selects candidates for an interview, the primary output is the candidates they select. However, important and relevant outcomes might include both the quality of those candidates and the fairness of the selection process. If that's the case, then this situation has two outcomes where the penalty may apply. That doesn't mean that every aspect of the output is a relevant outcome within the scope of the penalty. The specific presentation style of the recommendation may be different for an AI or human hiring manager but is not likely a relevant outcome.

Some outcomes are valuable in general but not relevant in many specific situations. Does it make sense to ask which piece of art you trust the most or which is more moral? Probably not. For artistic creations, outcomes may include authenticity, quality, or liking. Accuracy would only be an appropriate outcome if the art is supposed to represent something specific or follow some instructions, and speed would only be relevant if there was a deadline.

Likewise, sometimes labeled outcomes might be opposite in a general sense, but within a situation one is the relevant and meaningful outcome. For example, fair and unfair are opposites. Yet the penalty can be appropriately applied to either fairness or unfairness, depending on which one was relevant in a situation. If the situational outcome was *fair*, then the penalty means we would perceive it as less *fair* if produced by an AI. If the situational outcome was *unfair*, then the penalty means we would perceive it as less *unfair* if produced by an AI. If the situational outcome was neither fair nor unfair, then the penalty wouldn't apply as fairness—or unfairness—and would not be relevant in that situation. Throughout the book, I try to refer only to the outcome relevant in that situation, not its opposite, even though, by definition, less fairness is the same as more unfairness.

Taken all together indicates,

Relevant Outcomes Condition: The penalty applies to outcome(s) which are perceived to be relevant to a situation.

2.1.5 Similar Outcomes

The penalty applies to situations with *similar outcomes*.

First, this implies that if there are multiple important outcomes (*Relevant Outcomes Condition*) that all of them are similar. If an AI and human hiring manager were both similarly fair in process, but one always picked the worst candidates, then we'd prefer the one that did a better job. Conversely, if both ended up picking the same candidate in a given case, but we knew either the AI or human was highly sexist in their process, the non-sexist option would be superior. In either situation, the dissimilarity of one critical outcome means the penalty should not apply, assuming that aspect of the situation was being considered (*Relevant Outcomes Condition).*

Second, similar does not always mean identical. In some domains, AI and humans have a discrete number of choices: there are only so many candidates to interview and so many legal moves in chess. On the other hand, even a short paragraph on a topic can be written in millions of ways and the possibilities for creative compositions are innumerable. If one outcome is better, faster, cheaper, more detailed, or superior compared to the other, and that matters to us in this situation (*Relevant Outcomes Condition*), then the outcomes are inherently not similar. In that case, we will make our decision based on that difference, not the machine penalty.

Whether aspects of outcome are similar can only be determined within the context of the situation being evaluated (*Situational Effect* and *Relevant Outcomes Conditions*). If we're asked to evaluate two existing pieces of art, the speed with which it was originally created probably has no bearing, but if we need a specific illustration for an event later in the day, speed might be paramount, thereby rendering the decision dissimilar, assuming a human artist needed more than a few hours and AI need at most a few minutes.

Generally, outcomes that are more objective and quantitative are only considered similar when they are identical or nearly identical, whereas subjective and qualitative aspects tend to have a wider berth, where outcomes that are not identical could still be considered generally similar. Therefore, any difference in speed, price, quantity, accuracy, correctness, success, and winning can represent a meaningful difference in the outcome, whereas for trust, quality, liking, accuracy, genuineness, appropriateness, authenticity, personalization, fairness, morality, goodness, or potency, it takes a clearer, more substantive difference to consider the outcomes dissimilar. This is why the machine penalty will often be applied to quality, trust, and fairness, but less often to speed or price. Overall,

> *Similar Outcomes Condition: The penalty applies primarily to AI and human outcomes that are the same or not perceived as meaningfully different in a situation.*

In real-world situations, there are often a lot of dissimilarities in how AI and humans perform, create, decide, and behave, so it's worth discussing what this means in regard to this last condition. After all, if you listen to music composed by an AI followed by music composed by a human, you might consider the tempo, timbre, and melody to be *meaningfully different.* Or if you watch an autonomous vehicle on the road and notice the acceleration, braking patterns, distancing, and speed differ from the human population of drivers, you may also find this to be a *meaningful difference.* That could be enough for the penalty not to apply.

In contrast, major dissimilarities do not necessarily negate the application of the machine penalty if the differences are not meaningful to the situation and outcome. Perhaps you're comparing a human's photograph of nature with an AI's photorealistic alien world. Those are dissimilar in content, but if the images are similarly emotionally evocative, they may not be meaningfully different in terms of a penalty for emotional impact. Alternatively, you may not fully know the outcome, or notice all relevant aspects of it.

There is no doubt that in real-world situations there are many differences which are arguably meaningful. That is why much of the research that supports the machine penalty uses experiments. In experiments, the researchers are able to hold the meaningful differences constant and only change those things which are being studied. For example, using the exact same pieces of music and telling participants that they were created by either an AI or a human eliminates any dissimilarities based on the outcome itself. It leaves only one difference: who or what we believe

Table 2.1 Five conditions defining the scope of the machine penalty

Condition	The penalty applies...
Technology Entity Condition	...to any AI, machine, program, or other technology that is perceived as its own entity which can cause an outcome.
AI vs Human Condition	...when one believes some outcomes are caused by AI(s) versus human(s).
Situational Effect Condition	...across a range of situations, but some types of situations are expected to enhance, diminish, or even reverse the penalty.
Relevant Outcomes Condition	...to outcome(s) which are perceived to be relevant to a situation.
Similar Outcomes Condition	...primarily to AI and human outcomes that are the same or not perceived as meaningfully different in a situation.

created the music. Thereby, any differences in quality or enjoyment are linked to the creator, not the outcome, as the cause. For this reason, in the social sciences experiments are the gold standard to understand causation. In real-world situations, it will often be harder to peg a situation as 100% machine penalty, as alternative causes could be at play.

Before moving on to alternative causes, I'll recap the five formal conditions, which are summarized in Table 2.1.

2.2 The Cognitive Psychology of the Penalty

One such alternative cause is your own psychology, including the socio-cognitive processes such as attributions and biases. Social psychology is full of research on how we make inferences about people and their environment based on behavior, known generally as attribution processes and attribution theories.

For example, if a student fails a test, but has done well on other tests, we might attribute this to some external cause like the particular material or the student being ill. If a student fails several tests, we might focus the attribution on the student's ability, forming the impression they are either not studious or not smart. However, several different pieces of behavioral information could shift these attributions and the resulting impressions. If nearly all students did poorly on a test, then our attribution shifts back to the course material or perhaps how it's taught. If a single student did poorly on all exams, but did well on other assignments, that might shift the attribution to the student's test-related abilities and the test's format. Finally, if the student kept doing poorly on tests and assignments, but we found out they were going through difficult personal issues, again we might shift some of the attribution away from the student themselves and more on their overall life situation (Kelley, 1972; Malle et al., 2014; Shaver, 1985; Stryker & Gottlieb, 1981; Weiner, 1986).

This extended example shows how attributions from outcome to causation are complex, and how additional information can alter these attributions. We tend to make snap judgments and attributions when observing a behavior or seeing an outcome, and only when motivated enough to find or attend to other information do we change our judgment. This is one reason why first impressions are so important. It's also why we make poor judgments in the heat of the moment, only to regret them later. This contrast can be described as hot versus cold cognition (Salehinejad et al., 2021), fast versus slow thinking (Kahneman, 2011), or automatic system 1 thinking versus controlled and rational system 2 thinking (Evans, 2003).

Considering additional information or slowing and rationally considering one's attribution and impressions is an important sociocognitive process that will affect our judgments of AI, and therefore the application of the penalty. However, it could affect it in different ways including causing, eliminating, or superseding the penalty.

You might make a snap judgment about how wrong an outcome is. But then, on further reflection, you decide that because it came from an AI, there was no malicious intent behind it, and it was just an unfortunate result of the programming. If this leads you to then adjust your evaluation of how wrong it was, this two-step attribution and rational thinking process lead to the penalty.

Alternatively, you might immediately think that a humanlike chatbot is cheesy when it smiles at you and reacts to your jokes, while only on further reflection you decide that its reactions are both legitimate and meaningful. In this case, the slower thinking eliminated the penalty for the outcome of authenticity.

It's also possible that additional information might also supersede the penalty, making it no longer apply. If you are judging whether a human driver or the autonomous vehicle is responsible for an accident, the penalty suggests that you would blame the autonomous vehicle less than the human driver. However, on learning that the vehicle's tire was punctured by an undetectable nail on the road, your new attribution might remove blame and responsibility from both the human driver and the autonomous vehicle.

Finally, the attributions and thinking style have a clearer relationship with the penalty if someone is biased to make a specific attribution. For example, you can often find meaningful differences in specific situations if you desire to see the human's outcomes as greater than or less than the AI's.

Essentially, if you have prejudice against AI, you can usually search for reasons to perceive them as less than human such as finding minute differences in their outcome as your justification. But prejudice against AI can also work in lockstep with the penalty, doubling our reasons to diminish their outcomes. That is, initial preference for humans plus searching for a meaningful difference might, together, lead to diminishing AI.

2.3 Machine Prejudice Versus Machine Penalty

The penalty-as-prejudice idea suggests that we should consider whether this machine penalty is an ethical argument. One way to think about *comparing AI to humans leads us to diminish similar outcomes from AI across situations* is that we perceive AI as less than human and discriminate against them. I chose to use the loaded word *discriminate* here, considering how its multiple meanings (Harnois, 2023) applied to the machine penalty can help clarify it and its ethical implications.

Discriminate can mean either to differentiate between categories, to treat members of those categories differently, or to treat members of those categories prejudicially or unjustly. Thus, we can differentiate between AI and humans—the human–AI comparison—treating them differently for either valid or prejudicial reasons.

We frequently discriminate between humans in non-prejudicial, ethical ways most akin to the treat "differently" definition of the word. For example, we discriminate on who to hire for a job based on their qualifications and we discriminate on who can drive vehicles, with some classes of humans—like children—restricted from legally doing so, out of concerns about safety. These are forms of appropriate discrimination between groups and individuals, whereas hiring based on an applicant's race or not allowing women to get driver's licenses—as has been the case in some countries—are unjust, inappropriate, and unethical.

The machine penalty interestingly can encompass both types of discrimination. We might appropriately discriminate against ChatGPT writing an important memo, if we believe it could hallucinate and make up sources and we find the output of not making up sources relevant. The machine penalty suggests that the difference in perceived behavior in how ChatGPT writes the memo leads us to diminish it compared to a human writer—ourselves or a colleague—who will not make up sources. On the other hand, we might simply dislike ChatGPT or have prejudice against chatbots, and not use it based on that. Here the machine penalty is that we diminish machines, or ones with specific identities like chatbots, simply because they are a different, nonhuman outgroup. Likewise, as a society we may want to appropriately discriminate against autonomous vehicles if they aren't safe enough, although individually, people may fear them and decide to personally discriminate against using them.

That is where the paths of human and AI discrimination part ways. AI discrimination based on prejudice may be inaccurate—ChatGPT really might be able to help you and the autonomous vehicle may in fact be extremely safe. But prejudice leading to discrimination against AI, in my belief, is not morally wrong because AI are machines, not humans. In contrast, human discrimination based on prejudice may be inaccurate or unproductive as well—restricting qualified candidates based on race might lead to suboptimal hiring or restricting one gender from driving might lead to a less productive society. Differently from AI, human discrimination based on prejudice is always wrong based on the inherent value of humans.

Whether people exhibit the machine penalty for a good and valid reason or based on a more prejudicial attitude toward machines doesn't make it ethical or unethical. The machine penalty isn't harmful to AI, and, at this time, robots shouldn't be entitled to rights, an idea I return to in Chap. 13. But even amoral judgments which diminish the outcomes of AI may have ethical implications, depending on the domain and context of the judgments. We'll see this across five domains in Part II and in the downstream moral and societal costs for us as humans in Part III.

2.4 The Machine Advantage

When making a large sweeping argument, there are usually exceptions. While I believe there is a lot of evidence for the machine penalty, there is some evidence where it does not occur, and even some evidence that it's reversed. I'll call the reverse of it *the machine advantage*, which will be the case when *comparing AI to humans leads us to enhance similar outcomes from AI*. Overall, I think the scales are tipped toward the machine penalty, not the machine advantage. However, in specific situations, different causes of the machine penalty provide nuance to distinguish when the machine penalty or the machine advantage is more likely to apply (*Situational Effect Condition*).

Not only do I highlight cases of the machine advantage to keep a balanced, evidence-driven perspective, but it also contributes a lot to the understanding of the machine penalty.

When doing an experiment, the simplest part is examining differences between the averages of two different conditions. There is evidence for the hypothesis $X < Y$ if condition X's average is statistically less than condition Y's average. Likewise, the average trust in decisions made by AI being statistically less than average trust in the same decisions made by humans would be evidence for the machine penalty. However, the average X being less than the average Y does not mean that *all* Xs are less than *all* Ys, just like in some cases we might trust an AI's decision more than a human's.

Those cases that run counter to the general trend are actually quite telling for understanding the process by which something works. Similarly, situations that show evidence for the machine advantage probably shed more light on the machine penalty, because they reveal the conditions where it not only doesn't work but in fact is reversed.

But to understand the machine penalty or advantage, we first need to understand the causes, and to do that we need to consider how we perceive and diminish machines. We take up the research literature on perceiving machines next.

References

Evans, J. S. B. T. (2003). In two minds: Dual-process accounts of reasoning. *Trends in Cognitive Sciences, 7*(10), 454–459. https://doi.org/10.1016/j.tics.2003.08.012

Harnois, C. E. (2023). The multiple meanings of discrimination. *Social Psychology Quarterly, 86*(4), 413–431. https://doi.org/10.1177/01902725221134266

Kahneman, D. (2011). *Thinking, fast and slow*. Farrar, Straus and Giroux.

Kelley, H. H. (1972). Causal schemata and attribution process. In E. E. Jones, D. E. Kanouse, H. H. Kelley, R. E. Nisbett, S. Valins, & B. Weiner (Eds.), *Attribution: Perceiving the causes of behavior* (pp. 151–174). General Learning Press.

Malle, B. F., Guglielmo, S., & Monroe, A. E. (2014). A theory of blame. *Psychological Inquiry, 25*(2), 147–186. https://doi.org/10.1080/1047840X.2014.877340

Salehinejad, M. A., Ghanavati, E., Rashid, M. H. A., & Nitsche, M. A. (2021). Hot and cold executive functions in the brain: A prefrontal-cingular network. *Brain and Neuroscience Advances, 5*. https://doi.org/10.1177/23982128211007769

Shaver, K. G. (1985). *The attribution of blame: Causality, responsibility, and blameworthiness*. Springer-Verlag.

Stryker, S., & Gottlieb, A. (1981). Attribution theory and symbolic interactionism: A comparison. In J. H. Harvey, W. J. Ickes, & R. F. Kidd (Eds.), *New directions in attribution research* (Vol. 3, pp. 425–458). Erlbaum.

Weiner, B. (1986). *An attributional theory of motivation and emotion*. Springer-Verlag.

3

How Do We Perceive Machines?

To understand how we make human–AI comparisons, we first consider how we perceive machines in general, then in the next chapter how we perceive them differently compared to humans. Two social psychological perspectives specific to technology paved the way for understanding how we perceive machines. The first, *Computers Are Social Actors*, helps establish that we often automatically behave toward machines like we would behave toward humans whenever the machines provide humanlike signals. The second, anthropomorphism, shows that we imbue humanlike characteristics onto machines both based on characteristics of the machine and our own psychological needs.

3.1 *Computers Are Social Actors*

Byron Reeves and Clifford Nass in the late 1980s and early 1990s began a collaboration today known as *Computers Are Social Actors*, featured in their 1996 book: *The Media Equation: How People Treat Computers, Television, and New Media Like Real People and Places* (Reeves & Nass, 1996). Their argument was simple, but profound, especially for the time.

D. B. Shank, *The Machine Penalty*, https://doi.org/10.1007/978-3-031-86061-4_3

They argued and empirically validated that we act toward technology in psychologically similar ways as we act toward other humans. As long as the computer or other technology like a TV, character on a screen, or computerized voice inhabits a social role, behaves in a social manner, or has other social cues, people default to judging and interacting with it in a social psychological way (Nass, Steuer, & Tauber, 1994; Reeves & Nass, 1996).

The way they tested it was simple, but convincing. First, you find a well-established social psychological finding where a human interacts directly with or observes another human. It needs to be social, because their theory is all about human–technology social interaction.

Second, you create or adapt a simple and straightforward existing experiment, like one with two conditions. Perhaps in the original experiment a human participant would interact with a *human actor*, who unknown to the participant is working for the researcher and randomly assigned to perform either *behavior X* or *behavior Y*. Then the participant reacts to the *human actor*. Do this with enough participants and we can determine whether the participant's reaction is caused, or caused in part, by the difference in the *human actor's behaviors X* and *Y*. Since that's the basic experimental method in social psychology, there were plenty of experiments that could be used.

Third, Reeves and Nass would create a new experiment by replacing that *human actor* with a *computer*, or character or other technology. Sometimes it meant a slightly different format with a *computer* as one's interactant, but conceptually they made sure the key behaviors were still being performed. *Behavior X* and *behavior Y* should still be the same type of behaviors.

Fourth, if after running participants in this new experiment they found the same social psychological finding for this human–computer interaction as researchers had previously found for human–human interaction, then *Computers Are Social Actors* was supported. Computers were functioning as social actors.

3.1.1 Experimental Results

They did find, over and over, that computers are, in fact, treated as social actors, a profound breakthrough at the time.

When they studied politeness, they brought participants into the laboratory and had them interact with a computer which tutored them by telling them facts about Americans. Then the participants rated how well the tutoring computer did using a number of adjective descriptions. The two conditions involved half the participants answering these ratings using the computer that just tutored them and half using a different computer across the room. Just like we are more polite to people when we evaluate them face-to-face than when we evaluate them behind their backs, Nass and Reeves found that people rated the tutoring computer better when they did the rating directly on the tutoring computer compared to doing the rating on another computer (Nass, Steuer, & Tauber, 1994; Reeves & Nass, 1996).

In a twist on this experiment, they used a similar setup where a computer tutored them by telling them facts about Americans, which might be considered doing a good or bad job at tutoring. Then either the tutoring computer or another computer either praise or criticize how well the tutoring computer performed. In all cases, these conditions were randomly assigned, and the actual tutoring and feedback the human participant received were always identical. Yet, just as we like people more when others praise them compared to when they praise themselves, participants liked the tutoring computer more when it was praised by another computer compared to giving itself praise.

Similarly, participants disliked a computer more when it criticized another computer than when it criticized itself. But interestingly, in a more nuanced finding that also replicates human–human interaction, while criticizing another computer decreased likability, it also increased perceptions of intelligence (Nass, Steuer, Henriksen, et al., 1994; Reeves & Nass, 1996).

One of their most vivid experiments was how we perceive specialized experts. We believe that specialist human experts have superior knowledge in their domain compared to generalist human experts. For example, a *social psychologist* is presumed to know more about social psychology than someone who is known only as a *psychologist*, and not as a *social psychologist*.

Nass and Reeves brought people into the lab and had them all watch identical videos, which included both news stories and sitcom clips and then rate them. The only difference was that half the participants were randomly assigned to a generalist condition and half to a specialist condition. In the generalist condition, they watched all the clips on one television that was labeled "News and Entertainment Television." In the specialist condition, they watched the news stories on a television labeled "News Television" and the sitcom clips on an identical television next to it labeled "Entertainment Television."

This minor difference led participants to rate the news from the specialist news TV as more important, informative, interesting, and serious than the same news from the generalist TV. Similarly, they rated the sitcom clips as higher quality, funnier, more relaxing, and generally liked them more when presented on the specialist entertainment TV instead of the generalist TV (Nass, Reeves, & Leshner, 1996; Reeves & Nass, 1996).

In addition to politeness and expertise, Reeves and Nass found evidence for people treating technology like humans across a range of psychological processes (Reeves & Nass, 1996) including interpersonal distance (Reeves et al., 1992), flattery (Fogg & Nass, 1997), voice location (Nass & Steuer, 1993), reactions to size, sounds, and motion (Detenber & Reeves, 1996), personalities (Moon & Nass, 1996; Nass et al., 1995; Reeves & Greenberg, 1977), verbal and nonverbal cue consistency (Isbister & Nass, 2000), valence—whether something is good or bad—and arousal impressions (Detenber & Reeves, 1996; Reeves et al., 1989), emotion and memory (Newhagen & Reeves, 1992, 2013; Reeves et al., 1991), team processes (Nass, Fogg, & Moon, 1996), and gender stereotypes (Nass et al., 1997).

3.1.2 Expansions and Explanations

Over time there have been a number of expansions and revisions of the *Computers Are Social Actors* theory. For example, it's been applied to a wide range of modern technologies and media such as robots, smart phones, AI, chatbots, and websites (Gambino et al., 2020; Karr-Wisniewski & Prietula, 2010; Lombard & Xu, 2021; Xu et al., 2023), usually supporting the original hypothesis.

Interestingly, *Computers Are Social Actors* no longer works for a few of the original findings: like the one about politeness to desktop computers. Replicating the desktop politeness study, Evelien Heyselaar found there were no differences in how polite we are to a desktop computer who tutors us, when we're evaluating it on that desktop versus on another desktop (Heyselaar, 2023).

Why is this? Between the original study and the replication, a lot has changed: not only are desktops now old technology but people assume that many devices including computers are connected and share information. Therefore, perhaps we are less focused on the device itself and more on their current program or app. Therefore, as technology changes and humans' enculturation of the technology changes, researchers have proposed updates and new specifications of the theory (Gambino et al., 2020; Lombard & Xu, 2021; Xu et al., 2023).

One notable area of ongoing research is the mechanism by which *Computers Are Social Actors* works. By mechanism, I am referring to the answer to *Why?* or *What process causes people to interact with a computer like they interact with a human?* Originally, the primary explanation was one of mindlessness or automaticity which is when individuals automatically apply social rules, stereotypes, and expectations to technology (Nass & Moon, 2000). Mindlessness has been pitted against mindfulness—or a more active, intentional process—and the evidence tends to give more support to the mindlessness explanation (Kim & Sundar, 2012; Xu et al., 2022).

However, what specifically is mindlessly being processed has also been called into question. The idea of mindlessness in the original *Computers*

Are Social Actors paradigm argued we mindlessly apply scripts—that is, set routines and expectations—from human–human social interaction. But now we may mindlessly apply scripts from human–technology social interaction, given how often we interact with technology (Gambino et al., 2020).

Overall, *Computers Are Social Actors* has been supported by quite a bit of evidence, notwithstanding the studies that show it sometimes no longer applies. The support for it includes the mindlessness explanation, making it an important cornerstone in perceiving technology, and ultimately human–AI differences.

3.2 Anthropomorphism

The other major cornerstone, anthropomorphism, is the psychological process of perceiving humanlike attributes, emotions, motivations, and other mental states in nonhumans. Examples of it range from believing that your cat has missed you, that a tornado is vengefully selecting targets, or that your vehicle isn't starting because it's feeling depressed. An important consideration with anthropomorphizing machines, including robots, chatbots, devices, vehicles, or software, is that they have been explicitly designed. Some of their designs mimic or emulate humans in such ways that take advantage of our tendency to anthropomorphize.

The line between anthropomorphizing and accurately depicting a nonhuman is often pretty thin: you can logically describe your cat racing toward you when you put out their food as motivated by hunger, whereas assuming they desire the benefits of high-end pet resort pampering more accurately describes our human ideal of relaxation than theirs (Darling, 2021). Likewise, an android that furrows its brow and loudly tells you about a problem is certainly expressing anger like humans do, even if they are not experiencing the internal emotional state. Yet just because we anthropomorphize does not negate the potential for animals or machines to possess a mind, motivations, or thought processes that are similar to humans.

Therefore, anthropomorphism is more about our tendency as humans to see everything as human, than it is about the actual realities of the nonhuman world. We have a tendency to do this spontaneously with even very simple nonhumans. People spontaneously give human descriptions to both computer-animated blobs (Morewedge et al., 2007) and simple geometric shapes moving around (Heider & Simmel, 1944).

3.2.1 Why We Anthropomorphize

Nicholas Epley and colleagues (2007) have proposed three critical factors that explain why we anthropomorphize and thus are factors that increase our tendency to anthropomorphize.

First, it's based on our own human or self-knowledge—we see our own behaviors, attributions, emotions, and motivations as likely behaviors, attributions, emotions, and motivations for nonhumans as well, leading to the pet resorts with human-esque pampering. Therefore, we anthropomorphize AI and machines the more they are designed with human-like features (Zheng & Jarvenpaa, 2021) and the more they reflect ourselves (Alabed et al., 2022).

This means that as we move from interacting with computers and simple agents to robots and androids, people tend to increasingly anthropomorphize by attributing to the machine its own mental state (Thellman et al., 2022). Relatedly, people have stable differences in terms of the degree to which they anthropomorphize (Waytz, Cacioppo, et al., 2010), indicating that anthropomorphism is based in part on individual differences of the perceiver, in addition to features of the technology and situation.

The second factor is that we anthropomorphize to more effectively interact and accomplish more with the nonhuman. By predicting or explaining a nonhuman's behavior we can better understand how to best interact with it, or at least convince ourselves we do. When computers, robots, or machines act unpredictably or malfunction, people increasingly anthropomorphize them, rating them as having more intention, mind, free-will, and consciousness (Waytz, Morewedge, et al., 2010).

Thus, by anthropomorphizing, we create a more familiar, controlled situation where we are interacting with an intentional mind, rather than be puzzled by a malfunctioning and unpredictable machine. This process is more pronounced if the outcome of the machine's action is negative, as we spend more effort to find explanations for negative outcomes than we do for positive ones (Morewedge, 2009). Interestingly, even those who design or have knowledge of the internal working of the machines like robot designers, supercollider engineers, and car enthusiasts still succumb to this anthropomorphizing-to-explain phenomena, indicating that anthropomorphism is not curbed by superior knowledge of the machine (Epley, 2015, pp. 71–72; Shank et al., under review).

The third factor that increases anthropomorphism is making a social connection to the nonhuman. If the second factor is related to our thinking and cognition, this factor is related to our feelings, emotions, and relationships. People who are lonely tend to anthropomorphize more than those who aren't lonely (Bartz et al., 2016; Epley et al., 2008), partially because anthropomorphizing makes that nonhuman fulfill the need for social connection (Waytz et al., 2013) and reduces anxiety around social attachment (Bartz et al., 2016). People anthropomorphize in order to have social connections with a range of technologies including computers (Shin & Kim, 2020), mechanical gadgets (Epley et al., 2008), and robots (Shank et al., under review).

3.2.2 The Consequences of Anthropomorphism

Anthropomorphism itself leads to many critical consequences for both the person doing the anthropomorphizing as well as the object that is being anthropomorphized. In general, the more people anthropomorphize an object, the more they value and appreciate it (Chandler & Schwarz, 2010; Kwok et al., 2018). Anthropomorphizing robots and AI also increases how strongly we feel about them (Kim, Schmitt, & Thalmann, 2019; Shank, Graves, et al., 2019), trust them (de Visser et al., 2016; Waytz et al., 2014), empathize with them (Riek et al., 2009; Zhao & Malle, 2022), and want to have a continued relationship with them (Chiang et al., 2022; Koike et al., 2023).

Sometimes the outcomes of ongoing anthropomorphism can be fairly extreme. Soldiers have become upset when their unit's iRobot PackBot was destroyed by activating a bomb, holding a funeral and writing letters of commendation for it (Carpenter, 2016; Darling, 2021). Finally, anthropomorphizing robots can lead to some negative effects such as leading to increased dehumanization of other people, which is perceiving them as less human (Desbuleux & Fuss, 2023).

3.3 From *Computers Are Social Actors* to Anthropomorphism

It's worth noting that *Computers Are Social Actors* and anthropomorphism are not oppositional theories, nor does one supersede the other. Both can and do operate at the same time. Usually, disagreements between the two come at the level of the mechanism: either considering whether people are mindlessly or effortfully evaluating an AI or other entity, and whether people are simply responding to existing social cues or perceiving, and sometimes elaborating on, those social cues as representing humanness (Kim & Sundar, 2012; Nass & Moon, 2000; Xu et al., 2022).

One way to consider the relationship between the two is that *Computers Are Social Actors* is the first floor of a building and anthropomorphism is the second (see Fig. 3.1). If we did not respond to machines with any

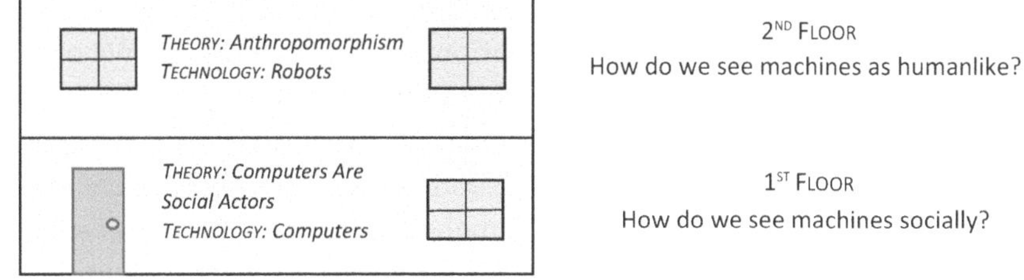

Fig. 3.1 The three floors of how we perceive machines representing the development of theories, the primary technology that each theory is applied to, and the primary question that each theory answers

type of social reactions like *Computers Are Social Actors* suggests we do, then it wouldn't make a difference if the machines were anthropomorphized to be more humanlike or not. For the first floor of the building, *Computers Are Social Actors* establishes that machines belong in the realm of the social and that they are legitimately perceived as social actors. For the second floor, anthropomorphism establishes that machines can be different types of social actors, some being more humanlike than others.

3.4 Perceiving Machines Compared to Perceiving Humans

But neither *Computers Are Social Actors* nor anthropomorphism fully builds up to the third floor—how machines directly compare to humans (Fig. 3.1). Most of the research in these foundational theories just focuses on the machine or other nonhuman and does not directly make the comparison to humans. This makes sense given the broad purposes of these theories: in general, we already know that humans are social actors, and we don't need to anthropomorphize humans in order to perceive them as humans. Of course, we do vary in how we perceive and treat humans, but these theories primarily relate to how we treat nonhumans, not comparing them to humans.

Therefore, this third floor is where the machine penalty is situated, and it is the situation we find ourselves more commonly dealing with today. If you found yourself in this chapter saying, *Of course we treat machines socially,* and, *Aren't most AI systems designed to look or act anthropomorphically?*, it's because, as a society, we're now living on the third story (see Yam et al., 2025 for a similar conceptualization of social machines). We often expect the other two floors and take them for granted.

This is not to say the other floors—and related theories—are not important. On the contrary, they are not only foundational, but we still encounter examples of them all the time. We might say "Thank you" to our smart home device for turning the lights on because we've asked vocally which feels like a social request, demonstrating the essence of *Computers Are Social Actors.* Additionally, when this smart home device is named Alexa and has a voice and physical presence, we're more likely to anthropomorphize it, attributing human motivations and cognitions to it. *Computers Are Social Actors* and anthropomorphism are still alive and well.

But now we have many machines that are acting like humans, have names like humans, voices like humans, and that can do tasks like humans. We generally call those machines *artificial intelligence*. Because of these machine capacities, our encounters—and perceptions of them—are often situated on the third floor, which is where we see the machine penalty.

While it's a bit of a simplification, the floors of this house and related theories apply most directly to different types of machines (Fig 3.1). *Computers Are Social Actors* most directly applies to computers because computers can serve many of the same roles as humans. Anthropomorphism most directly applies to robots and androids because they are designed to appear like humans and physically mimic human functions. The machine penalty most directly applies to artificial intelligence as AI's complex decision-making puts it in competition and comparison with humans' decisions and output. The theories, of course, can be applied to other technologies, but delegating a primary machine to each theory is a useful way to distinguish their purposes.

A final way to think about the different floors is what question each theory answers (Fig. 3.1). *Computers Are Social Actors* fundamentally answers, *How do we see machines socially?* Anthropomorphism answers, *How do we see machines as humanlike?* The machine penalty answers, *How do we see machines compared to humans?* The answers, as we explore in the next three chapters, are that we tend to compare their differences (Chap. 4), and then diminish them (Chap. 5) based on a number of causes (Chap. 6).

References

Alabed, A., Javornik, A., & Gregory-Smith, D. (2022). AI anthropomorphism and its effect on users' self-congruence and self–AI integration: A theoretical framework and research agenda. *Technological Forecasting and Social Change, 182*, 121786. https://doi.org/10.1016/j.techfore.2022.121786

Bartz, J. A., Tchalova, K., & Fenerci, C. (2016). Reminders of social connection can attenuate anthropomorphism: A replication and extension of Epley, Akalis, Waytz, and Cacioppo (2008). *Psychological Science, 27*(12), 1644–1650. https://doi.org/10.1177/0956797616668510

Carpenter, J. (2016). *Culture and human-robot interaction in militarized spaces: A war story*. Routledge.

Chandler, J., & Schwarz, N. (2010). Use does not wear ragged the fabric of friendship: Thinking of objects as alive makes people less willing to replace them. *Journal of Consumer Psychology, 20*(2), 138–145. https://doi.org/10.1016/j.jcps.2009.12.008

Chiang, A.-H., Trimi, S., & Lo, Y.-J. (2022). Emotion and service quality of anthropomorphic robots. *Technological Forecasting and Social Change, 177*, 121550. https://doi.org/10.1016/j.techfore.2022.121550

Darling, K. (2021). *The new breed: what our history with animals reveals about our future with robots*. Henry Holt and Company.

de Visser, E. J., Monfort, S. S., McKendrick, R., Smith, M. A. B., McKnight, P. E., Krueger, F., & Parasuraman, R. (2016). Almost human: Anthropomorphism increases trust resilience in cognitive agents. *Journal of Experimental Psychology: Applied, 22*(3), 331–349. https://doi.org/10.1037/xap0000092

Desbuleux, J. C., & Fuss, J. (2023). Is the anthropomorphization of sex dolls associated with objectification and hostility toward women? A mixed method study among doll users. *The Journal of Sex Research, 60*(2), 206–220. https://doi.org/10.1080/00224499.2022.2103071

Detenber, B. H., & Reeves, B. (1996). A bio-informational theory of emotion: Motion and image size effects on viewers. *Journal of Communication, 46*(3), 66–84. https://doi.org/10.1111/j.1460-2466.1996.tb01489.x

Epley, N. (2015). *Mindwise: Why we misunderstand what others think, believe, feel, and want*. Vintage.

Epley, N., Akalis, S., Waytz, A., & Cacioppo, J. T. (2008). Creating social connection through inferential reproduction: Loneliness and perceived agency in gadgets, gods, and greyhounds. *Psychological Science, 19*(2), 114–120. https://doi.org/10.1111/j.1467-9280.2008.02056.x

Epley, N., Waytz, A., & Cacioppo, J. T. (2007). On seeing human: A three-factor theory of anthropomorphism. *Psychological Review, 114*(4), 864–886. https://doi.org/10.1037/0033-295X.114.4.864

Fogg, B. J., & Nass, C. (1997). Silicon sycophants: The effects of computers that flatter. *International Journal of Human-Computer Studies, 46*(5), 551–561. https://doi.org/10.1006/ijhc.1996.0104

Gambino, A., Fox, J., & Ratan, R. A. (2020). Building a stronger CASA: Extending the computers are social actors paradigm. *Human-Machine Communication, 1*, 71–85. https://doi.org/10.30658/hmc.1.5.

Heider, F., & Simmel, M. (1944). An experimental study of apparent behavior. *The American Journal of Psychology, 57*(2), 243–259. https://doi.org/10.2307/1416950

Heyselaar, E. (2023). The CASA theory no longer applies to desktop computers. *Scientific Reports, 13*(1), 19693. https://doi.org/10.1038/s41598-023-46527-9

Isbister, K., & Nass, C. (2000). Consistency of personality in interactive characters: Verbal cues, non-verbal cues, and user characteristics. *International Journal of Human-Computer Studies, 53*(2), 251–267. https://doi.org/10.1006/ijhc.2000.0368

Karr-Wisniewski, P., & Prietula, M. (2010). CASA, WASA, and the dimensions of us. *Computers in Human Behavior, 26*(6), 1761–1771. https://doi.org/10.1016/j.chb.2010.07.003

Kim, S. Y., Schmitt, B. H., & Thalmann, N. M. (2019). Eliza in the uncanny valley: Anthropomorphizing consumer robots increases their perceived warmth but decreases liking. *Marketing Letters, 30*(1), 1–12. https://doi.org/10.1007/s11002-019-09485-9

Kim, Y., & Sundar, S. S. (2012). Anthropomorphism of computers: Is it mindful or mindless? *Computers in Human Behavior, 28*, 241–250. https://doi.org/10.1016/j.chb.2011.09.006

Koike, M., Loughnan, S., & Stanton, S. C. (2023). Virtually in love: The role of anthropomorphism in virtual romantic relationships. *British Journal of Social Psychology, 62*(1), 600–616. https://doi.org/10.1111/bjso.12564

Kwok, C., Grisham, J. R., & Norberg, M. M. (2018). Object attachment: Humanness increases sentimental and instrumental values. *Journal of Behavioral Addictions, 7*(4), 1132–1142. https://doi.org/10.1556/2006.7.2018.98

Lombard, M., & Xu, K. (2021). Social responses to media technologies in the 21st century: The media are social actors paradigm. *Human-Machine Communication, 2*, 29–55. https://doi.org/10.30658/hmc.2.2

Moon, Y., & Nass, C. (1996). How "Real" are computer personalities?: Psychological responses to personality types in human-computer interaction. *Communication Research, 23*(6), 651–674. https://doi.org/10.1177/009365096023006002

Morewedge, C. K. (2009). Negativity bias in attribution of external agency. *Journal of Experimental Psychology: General, 138*(4), 535–545. https://doi.org/10.1037/a0016796

Morewedge, C. K., Preston, J., & Wegner, D. M. (2007). Timescale bias in the attribution of mind. *Journal of Personality and Social Psychology, 93*(1), 1–11. https://doi.org/10.1037/0022-3514.93.1.1

Nass, C., Fogg, B. J., & Moon, Y. (1996). Can computers be teammates? *International Journal of Human-Computer Studies, 45*, 669–678. https://doi.org/10.1006/ijhc.1996.0073

Nass, C., & Moon, Y. (2000). Machines and mindlessness: Social responses to computers. *Journal of Social Issues, 56*(1), 81–103. https://doi.org/10.1111/0022-4537.00153

Nass, C., Moon, Y., Fogg, B. J., & Reeves, B. (1995). Can computer personalities be human personalities? *International Journal of Human-Computer Studies, 43*(2), 223–239. https://doi.org/10.1006/ijhc.1995.1042

Nass, C., Moon, Y., & Green, N. (1997). Are machines gender neutral? Gender-stereotypic responses to computers with voices. *Journal of Applied Social Psychology, 27*(10), 864–876. https://doi.org/10.1111/j.1559-1816.1997.tb00275.x

Nass, C., Reeves, B., & Leshner, G. (1996). Technology and roles: A tale of two TVs. *Journal of Communication, 46*(2), 121–128. https://doi.org/10.1111/j.1460-2466.1996.tb01477.x

Nass, C., & Steuer, J. (1993). Voices, boxes, and sources of message: Computers and social actors. *Human Communication Research, 19*(4), 504–527. https://doi.org/10.1111/j.1468-2958.1993.tb00311.x

Nass, C., Steuer, J., Henriksen, L., & Dryer, D. C. (1994). Machines, social attributions, and ethopoeia: Performance assessments of computers subsequent to "self-" or "other-" evaluations. *International Journal of Human-Computer Studies, 40*(3), 543–559. https://doi.org/10.1006/ijhc.1994.1025

Nass, C., Steuer, J., & Tauber, E. R. (1994). Computers are social actors. SIGCHI Conference on Human Factors in Computing Systems: Celebrating Interdependence, Boston, MA.

Newhagen, J. E., & Reeves, B. (1992). The evening's bad news: Effects of compelling negative television news images on memory. *Journal of Communication, 42*(2), 25–41. https://doi.org/10.1111/j.1460-2466.1992.tb00776.x

Newhagen, J. E., & Reeves, B. (2013). Emotion and memory responses for negative political advertising: A study of television commercials used in the 1988 presidential election. In *Television and Political Advertising* (pp. 197–220). Routledge.

Reeves, B., & Greenberg, B. S. (1977). Children's perceptions of television characters. *Human Communication Research, 3*(2), 113–127. https://doi.org/10.1111/j.1468-2958.1977.tb00510.x

Reeves, B., Lang, A., Thorson, E., & Rothschild, M. (1989). Emotional television scenes and hemispheric specialization. *Human Communication Research, 15*(4), 493–508. https://doi.org/10.1111/j.1468-2958.1989.tb00196.x

Reeves, B., Lombard, M., & Melwani, G. (1992). *Faces on the screen: Pictures or natural experience*. International Communication Association.

Reeves, B., & Nass, C. (1996). *The media equation: How people treat computers, television, and new media like real people and places*. Cambridge University.

Reeves, B. R., Newhagen, J., Maibach, E., Basil, M., & Kurz, K. (1991). Negative and positive television messages: Effects of message type and context on attention and memory. *American Behavioral Scientist, 34*(6), 679–694. https://doi.org/10.1177/0002764291034006006

Riek, L. D., Rabinowitch, T.-C., Chakrabarti, B., & Robinson, P. (2009). Empathizing with robots: Fellow feeling along the anthropomorphic spectrum. 3rd International Conference on Affective Computing and Intelligent Interaction and Workshops, Amsterdam, Netherlands.

Shank, D. B., Graves, C., Gott, A., Gamez, P., & Rodriguez, S. (2019). Feeling our way to machine minds: People's emotions when perceiving mind in artificial intelligence. *Computers in Human Behavior, 98*, 256–266. https://doi.org/10.1016/j.chb.2019.04.001

Shank, D. B., Sunkpal, F., Wright, D., Krueger, M. A., Young, N.-L., & Gordon, K. (under review). Warrior Bot, Relational Bot, Blameworthy Bot: How Combat Robot Designers Anthropomorphize Their Robot Creations.

Shin, H. I., & Kim, J. (2020). My computer is more thoughtful than you: Loneliness, anthropomorphism and dehumanization. *Current Psychology, 39*, 445–453. https://doi.org/10.1007/s12144-018-9975-7

Thellman, S., de Graaf, M., & Ziemke, T. (2022). Mental state attribution to robots: A systematic review of conceptions, methods, and findings. *ACM Transactions of Human-Robot Interaction, 11*(4), 1–51. https://doi.org/10.1145/3526112

Waytz, A., Cacioppo, J., & Epley, N. (2010). Who sees human? The stability and importance of individual differences in anthropomorphism. *Perspectives on Psychological Science, 5*(3), 219–232. https://doi.org/10.1177/1745691610369336

Waytz, A., Epley, N., Cacioppo, J. T., & Akalis, S. A. (2013). Social connection and seeing human. In C. N. Dewall (Ed.), *The Oxford handbook of social exclusion* (pp. 251–256). Oxford University Press.

Waytz, A., Heafner, J., & Epley, N. (2014). The mind in the machine: Anthropomorphism increases trust in an autonomous vehicle. *Journal of*

Experimental Social Psychology, 52, 113–117. https://doi.org/10.1016/j.jesp.2014.01.005

Waytz, A., Morewedge, C. K., Epley, N., Monteleone, G., Gao, J.-H., & Cacioppo, J. T. (2010). Making sense by making sentient: Effectance motivation increases anthropomorphism. *Journal of Personality and Social Psychology, 99*(3), 410–435. https://doi.org/10.1037/a0020240

Xu, K., Chen, X., & Huang, L. (2022). Deep mind in social responses to technologies: A new approach to explaining the computers are social actors phenomena. *Computers in Human Behavior, 134*, 107321. https://doi.org/10.1016/j.chb.2022.107321

Xu, K., Liu, F., Chen, X., & Lombard, M. (2023). The Media Are Social Actors paradigm and beyond: Theory, evidence, and future research. In *Research Handbook on Artificial Intelligence and Communication* (pp. 337–353). Edward Elgar Publishing. https://doi.org/10.4337/9781803920306.00031

Yam, K. C., Eng, A., & Gray, K. (2025). Machine replacement: A mind-role fit perspective. *Annual Review of Organizational Psychology and Organizational Behavior, 12*. https://doi.org/10.1146/annurev-orgpsych-030223-044504

Zhao, X., & Malle, B. F. (2022). Spontaneous perspective taking toward robots: The unique impact of humanlike appearance. *Cognition, 224*, 105076. https://doi.org/10.1016/j.cognition.2022.105076

Zheng, J. F., & Jarvenpaa, S. (2021). Thinking technology as human: Affordances, technology features, and egocentric biases in technology anthropomorphism. *Journal of the Association for Information Systems, 22*(5), 1429–1453. https://doi.org/10.17705/1jais.00698

4

Human–AI Comparisons

Now that I've discussed some social psychological background on how we perceive machines in general, I want to compare how we perceive humans versus AI. This "third floor" question (Fig. 3.1) is really the beginning of the machine penalty process (Figs. 1.1 and 1.2) because these comparisons are the root cause of the penalty. We don't immediately think outcomes from AI are worse than those from humans, but instead we diminish them based on the comparisons we've already made between humans and AI. And to consider how critical and complex these comparisons can be to our evaluations, we'll start with the simpler issue of how we choose a slice of cake.

4.1 How Do We Choose Cake?

We humans don't just behave differently toward objects, technology, and people in the world for no reason: we react based on comparing aspects that differentiate these stimuli in our environment. Imagine you're at a bakery or store and selecting a piece of cake to buy. What determines your behavior? It probably has something to do with the appearance of

D. B. Shank, *The Machine Penalty*, https://doi.org/10.1007/978-3-031-86061-4_4

the cake—which one looks the best—and possibly which price tag looks the best too. You might have observed other aspects as well, like which one was baked most recently, or which cake is down to the final piece. Different decisions you make are based on your observations.

On the other hand, it would be silly to think your personal beliefs don't play a role. Most critically, what flavors you like should heavily influence your decision. But perhaps other beliefs about which cake will pair well with your coffee or whether you fancy yourself a cake connoisseur will also influence your decision. So, unless you're too hungry and have stopped reading to grab a snack, we might broadly say that both observations and beliefs combined to help us evaluate the slice of cake. And this is much more applicable for evaluations of complex stimuli like AI and machines, as well as other humans.

I've identified five meaningful comparisons we make between AI and humans: appearance, identity, behavior, mind, and essence. In this chapter, I discuss how these comparisons differentiate humans and AI, then in Chap. 6 return to evidence that they can cause the machine penalty. These comparisons range in terms of their basis in observation versus beliefs, as shown in Fig. 4.1. And like choosing cake, they aren't mutually exclusive but can operate simultaneously. For AI, we might be influenced by a cute robotic toy's appearance, but also consider that it doesn't really have a mind. We might also consider an AI recommender's record of success—that is its previous behavior—but that could be influenced by our essentialist belief that an AI can't really understand our personal decisions.

Another aspect that varies with observation versus belief is whether the human–AI difference applies primarily to an individual AI, more generally to other similar AI entities, or even collectively to all machines (Fig. 4.1). The observation that the carrot cake with cream cheese frosting was cut into small pieces and overbaked will influence you to not select it, but probably won't affect your evaluation of the *tres leches* cake or other carrot cakes in the future. If your belief is that carrot cake is the worst cake in the world, or that cake in general is just an inferior substitute, only good for when cookies and ice cream aren't available, then your belief will affect a large range of cake decisions.

Similarly, a robot with a human-like face—part of appearance—alters how we perceive and interact with that particular robot, but not others.

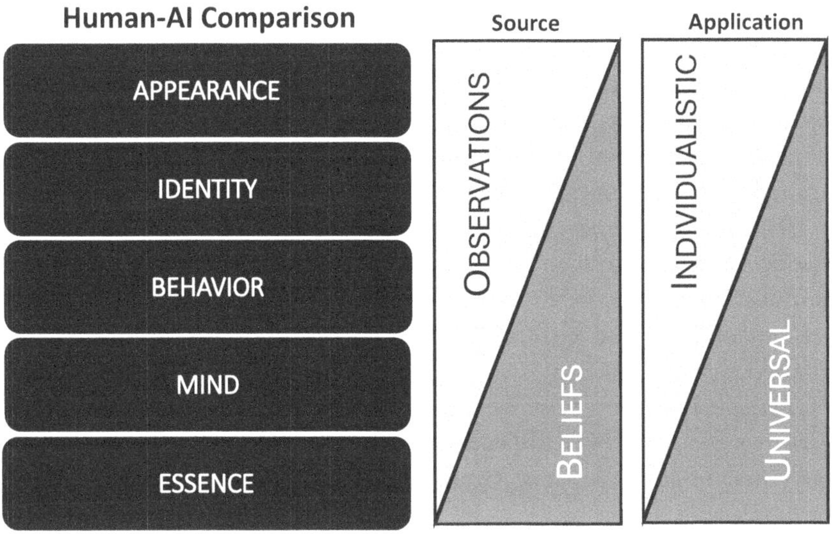

Fig. 4.1 Types of human–AI comparison based on their relation to observations or beliefs as sources of the information and application to individual machines or universally to all machines

On the other hand, if we believe that AI can never truly experience love—a belief about the essence of machines—then that belief is going to apply to every AI, robot, and machine you encounter. These comparisons are useful not only for differentiating between AI and humans but also can differentiate between different machines or between different humans.

4.2 Appearance

When you meet people, unless it's through text-only online communication, you subconsciously have an idea of their general age, their gender, skin tone, race, accent if they've spoken, and their specific look, including attractiveness and body type. You also, for a given encounter, may know more transitory appearance details like clothing, accessories, hairdos, tans, or an expression on their face that suggests too much caffeine or not enough sleep.

From this we make inferences and judgments. A middle-aged professionally dressed woman on campus is likely to be a professor. A young person with a muscular and fit body is probably better at sports than an older, less athletic person. And by listening for a particular accent along with skin tone, facial features, and even clothing, we can sometimes be accurate in determining someone's country of origin.

Yet, stereotypes and cognitive shortcuts often don't work. Young professors often get mistaken for students, while non-traditional students are mistaken for professors. Some of the best racquetball players are well past retirement age. And cues about race and country of origin are fluid enough that it's sometimes easy to shift others' beliefs (Khanna, 2010). Nonetheless, we form impressions, inferences, and even stereotypes about other people and groups of people. Not surprisingly, appearances are fundamental to how we view AI and robots and how we compare them to humans.

From PARO, the fluffy therapeutic robot seal to advanced military bomb diffusers, we find that robots come in all shapes and sizes and their physical appearance is often our first encounter with them. Non-robotic AI such as chatbots, NPCs in games, and recommender systems, may also have virtual bodies, a virtual head, or some other visual representation. Then again, some have no body, face, or representation.

Across all these forms, the physical or virtually represented appearance, or lack thereof, makes a difference in how we perceive these machines as not-quite human. First and most obviously, their appearance allows us to infer what physical acts will be possible or not possible. If a robot does not have a hand, it's not going to be able to hold your hand, and if it doesn't have wheels, legs, or similar, it will not be able to navigate through your house. More substantially, the appearance of human-like features plays into people's tendency to anthropomorphize, and different appearance cues trigger social responses, as would be expected from the *Computers Are Social Actors* paradigm. This suggests that appearance, especially human-like features, should make a big difference in how we perceive AI and robots.

And they do. Researchers who wanted to understand how people process humanlike appearance in robots cataloged a database of over 250 robots, identifying 16 humanlike features such as apparel, face, and gender, and statistically analyzed people's rating of them to find major dimensions of robot appearance: body-manipulators, face, surface look, and mechanical locomotion (Phillips et al., 2018; Zhao et al., 2019). Connecting to the differences in perceived mental capacity, body-manipulators like arms and legs increase the perception of agentic capabilities, while facial and surface features alter perception of emotional capacity (Zhao et al., 2019).

Faces also change how we view robots in more nuanced ways. People find robots a lot more trustworthy, credible, and honest when their faces are more baby-like (Song et al., 2021). Similarly, humans are more likely to take the perspective of a robot when the robot has a face with eyes, even if the robot is not highly human-like (Zhao & Malle, 2022). Voices, too, are an important part of how we judge robots and AI. Voices help humanize machines, making them seem more like individuals with their own minds (Schroeder & Epley, 2016; Schroeder et al., 2017). AI with anthropomorphic designs, which have more humanlike features, lead people to trust them more compared to those without such features (Waytz et al., 2014).

Therefore, with our tendency to anthropomorphize and treat computers as social actors, those robots, characters, and AI with humanlike physical features tend to be seen as closer to humans and treated as such. But there are exceptions, and the biggest exception is the uncanny valley—machines that seem fairly, but not fully, humanlike and end up coming across as creepy (Mori, 1970).

This can be a nearly perfect android with awkward movements (Thepsoonthorn et al., 2021), robot faces that lack realism (Kim et al., 2022; Mathur & Reichling, 2016), animated characters with imperfect CGI movements like in *The Polar Express* movie (Palomäki et al., 2018), game NPCs and avatars (Stein & Ohler, 2017; Tinwell, 2014), and not-quite-real animals (Löffler et al., 2020; Schwind et al., 2018). While research suggests a number of potential explanations for this phenomenon (Wang et al., 2015), one of the strongest is that of a perceptual mismatch, wherein some characteristics of the entity are humanlike and others artificial (Carr et al., 2017; Kätsyri et al., 2015; Mathur et al., 2020).

However, other lines of research suggest the process may extend beyond a visual process and also deal with our perception of the machine's mental capacities. For example, people find machines uncannier when they are autonomous AI that can demonstrate empathy (Stein & Ohler, 2017) or when they experience human sensations like hunger and fear (Gray & Wegner, 2012) and emotions (Appel et al., 2020). When I asked people to recall personal encounters with AI where they perceived it to have mental capacities like thinking and feeling, 19 out of 266 spontaneously mentioned uncanny feelings like unease or being creeped out in their encounter (Shank, Graves, et al., Shank, Graves, et al., 2019). Yet, a meta-analysis and experiment suggest that appearance matters more to uncanny perceptions than perceptions of mind (MacDorman, 2024).

Uncanny valley studies show appearance and other attributes can combine to alter our impression, illustrating that human–AI differences often overlap in the real world. Overall, AI appearance is critical for us to see practical limitations of an AI, to judge its human-likeness, and to help us determine AI's identity, behavior, and mind.

4.3 Identity

Like appearance, we encounter both AI and humans initially through their identities such as *Alexa* or *personal assistant*, social roles that fit with those identities like answering questions or adding to one's calendar, and even the specific labels we use for those identities, such as the difference between a *personal assistant* and a *personal secretary*. For humans, these identities, roles, and labels affect how we behave, what others think of us, and how others behave toward us (Burke & Stets, 2009; Heise, 2007; MacKinnon & Heise, 2010; Owens et al., 2010). One of the first things we learn about people is their name and usually other important identity information, like their occupation. Even for those we don't know, appearance and behavioral clues quickly help us navigate who in a store is an employee or customer, or who is in charge of a meeting.

When AI systems are created, they are often given identifying information, or at the least are identified with the company that created them. That's one difference between machine and human identities: the machine might be an Amazon, Apple, or Google machine, whereas a person could only identify with those companies as an employee, customer, or fan.

They may, like humans, also have a personal name, like Siri or Alexa, although there is a difference here as well: their personal name is shared by millions of other identically made Siris or Alexas.

Aside from these company and personal names, we might generally identify them as personal assistants. We might expect more versatile roles with the identity *personal assistant* than the identity *smart home assistant*, which limits the expectations to the home domain and differentiate them from humans who don't have that identity. We may also perceive machines based on both explicit cues, such as personal assistants often being designed with female voices and names (Feine et al., 2020), but also implicitly decide that our personal assistant is female, which many people do with or without explicit identity information (Fortunati et al., 2022).

AI identities often suggest more constraints than human identities do, because many AI are single-function only. A chess-playing AI likely can't play any other games, let alone paint your house, buy items on eBay, or mow your lawn. You wouldn't assume a human mowing your lawn only had one identity; he might also be a student, Minecraft player, and brother, among others. In this way, AI gives a stronger signal of either their expertise or inappropriateness, based on their identity. A machine that is labeled *Terra: iRobot's Robot Lawn Mower* has indicated its appropriateness in one domain and inappropriateness in all others. Your neighborhood teenage boy might also signal his expertise with fliers telling you about *Kevin's Lawn Mowing Services* precisely because humans inhabit multiple identities and roles. Kevin must have had some motivation and training to learn how to mow lawns, even if it was the minimal, *Hey, I could push the mower around and make money!* In contrast, Terra was designed to mow lawns and that's its only function.

Yet there are more general AI—known technically as artificial general intelligence—that are created with skills so general, that may enact many identities, more similar to humans. Advanced chatbots like ChatGPT are approaching this with broad language-based functions such as being able to write, read, and research. They are still limited to primarily language-and-communication-based tasks but aren't relegated to a specific identity. That's why people use them for a range of purposes such as summarizing articles, debating philosophy, providing feedback on recipes, or even companionship.

In these cases, the identities are not predefined, and people can switch from one to another. If you desired ChatGPT to enact the identity of a sympathetic friend, ChatGPT might only fulfill some roles typically related to that human identity. It might give acknowledgments of what you said and the emotions you are conveying, but it may too quickly offer a range of advice that it has absorbed. Then again, a sympathetic human friend might also give more advice than is desired. Connecting back to appearance, ChatGPT's lack of physicality means it would be unlikely to enact the role in ways that require a physical presence, such as putting a hand on your shoulder or taking you on a surprise outing to cheer you up.

Computers Are Social Actors research also shows that machines are perceived according to their identities, and the role expectations associated with them. This is why the specialist television is perceived as superior to the generalist television when delivering content related to its specialization (Nass, Reeves, et al., Nass, Reeves, & Leshner, 1996). But of course, that doesn't mean that a specialist television, or AI, is perceived exactly like a specialist human.

In many cases, AI and humans are perceived differently, even when occupying the exact same identity. For example, people blamed autonomous AI military drones at different rates than human military drone pilots for going ahead with or aborting a strike on terrorists when the situation changed, and a civilian child could be hurt. While participants in the study disagreed about the best ethical decision and their beliefs affected overall blame, people know soldiers are part of a command hierarchy in the military, whereas machines are not. Therefore, the identity of a human soldier changed the role expectation of whether they were free to make the choice themselves without explicit approval (Malle et al., 2019).

Identities also connect individuals to larger groups and even affect our evaluation of those groups. In one project, people saw immoral behavior by a chatbot or human assistant as indicating not only that the chatbot or human assistant, respectively, was likely to continue to behave immorally but also that all chatbots and all human assistants, respectively, were more likely to behave immorally. Yet, while an AI's immoral behavior did affect the likelihood of immoral behavior for AI as a group, immoral behavior of a human did not affect the likelihood of immoral behavior for humans as a group. Thus, this moral spillover to other related identities required more specific human identification, compared to an AI which only required the identification as an AI (Manoli et al., 2025).

Therefore, identities can form critical human–AI comparisons, even as humans and AI inhabit many of the same identities. AI are often less malleable in their identities, and we assume differences in motivations and experiences based on these identities, causing a large rift between humans and AI. They also set up expectations and constraints for differences in behavior.

4.4 Behavior

Like appearance and identities, there are the obvious behavioral comparisons that give away how AI are different from humans. This different behavior could be a robot that can't climb stairs or can pick up massive weights, or a chatbot that analyzes thousands of sources to answer the question you posed seconds ago.

Often AI behaviors reveal different capacities from humans. When AlphaGo played and beat Go Grandmaster Lee Sedol in a set of historic matches in 2016, in match 2 on move 37, it made "a creative move … [not] seen in a top player's game" described as "something new and unique [Sedol] has to think about" (Metz, 2016). This behavior not only showcased the AI's ability to master Go but also—as was noted by this commentator—showcased just how nonhuman AlphaGo seems. Since this AI only had limited behavioral options—legal Go moves—this example shows that even behaviors within constraints can effectively distinguish AI from humans.

Yet particular behavior comparisons aren't always positive for AI. For example, one medical diagnosis AI system recommended that patients with both pneumonia *and* asthma receive less intensive care than those with only pneumonia *or* only asthma. This was because it was trained on data that showed better final outcomes from those with both. This was due to patients' admittance to intensive care when they had both, even though the AI system wasn't trained on that information. It just logically concluded they needed less care because they had better final outcomes

(Caruana et al., 2015). Many times, this stark lack of common sense in behaviors serves to remind us that AI don't behave like humans.

Since the inception of artificial intelligence, the standard for determining whether AI was actually intelligent was its ability to imitate humans well enough to pass as one—a test of behavioral similarity. The Turing Test as it's now known involves human judges engaging in a text-based conversation with either an AI or human with the goal of trying to identify them correctly (Turing, 1950). The fundamental idea is that if computer programs can, through conversation, behave enough like humans to be mistaken for one, they are in fact intelligent. Intelligence, therefore, is based on understanding human behavior enough to mimic it, while simply answering questions faster with better spelling and grammar actually detracts from that purpose.

Whether the Turing Test has ever been passed depends on which rules (Epstein et al., 2008) and which criteria for passing—both subjects of much debate—are being used (Neufeld & Sonje, 2020). However, the concept of the Turing Test does suggest two issues about behavior as being an important part of the machine penalty process.

First, in cases where the Turing Test is passed in everyday life, meaning we aren't sure if we're interacting with an AI or human, the machine penalty no longer applies (*AI vs Human Condition*). If you can't tell if the business representative on the other end of your call is human or AI, or if you don't know whether music was composed by an AI or human, then you have no basis for comparison.

Second, the Turing Test illuminates the more fundamental differences between humans and AI, such that people who are trying to convince the judges they are human must decide themselves what is key to being human, or at least what others believe is key. One such participant found individual personality, changing topics, jokes, and non-sequiturs as key behaviors in order to be perceived as human by other humans (Christian, 2011).

Third, language is one key form of behavior. In the classic Turing Test and most versions of it the only interaction a judge has with the potential AI or potential human is through a text-only chat. Language is a key behavior that provides social and psychological information about other people (Kinzler, 2021), and is extremely important as the computational and interaction method of large language models like ChatGPT.

In a strange but fascinating experiment, researchers asked what one word would people say to convince someone they are human, not an AI? In this one-word Turing Test, people selected words relating to love and emotions, humans, politeness, humanness, mind, faith, nonhumans, life and death, bodily functions, and profanities. Using specific words from each category, the researchers then found that bodily functions won, with *poop* being the best option, while *love* and *please* were close seconds (McCoy & Ullman, 2018).

This research and the Turing Test showcase that their behaviors are an important aspect to make comparisons, and we have some intuitive ideas of what behaviors to expect from a machine compared to a human. And often that comes from our belief that AI have different types of minds.

4.5 Mind

Whether brains versus CPU, learning versus programming, or electronic versus biological, we often think about humans and machines in terms of their capabilities and the structures that cause those, and how they are similar or different. Roughly speaking, biologists and neuroscientists study the *brain*, whereas psychologists and cognitive scientists study the *mind*. The mind is often defined as one's thinking and feeling, the conscious awareness of that thinking and feeling, and cognitive processes such as focus, reaction, memory, distraction, bias, and sensory perception. Being aware of our own emotions and thinking makes it easier to attempt to understand the minds of others, including what they are feeling and thinking (Epley, 2015).

This is where social psychologists get interested—how do we guess what others are thinking and feeling? What do we assume about their mind? How does this influence our interactions with and relationships with others? For example, if someone makes an emotional plea, it makes sense to assume they have strong feelings about that issue. But our inferences about their mind change based on the social situation and their behavior. If your stoic and logical friend makes this plea, you might assume it's truly important to them, whereas if a politician has a soundbite of the same, you might discount it as pandering.

One popular area of perceiving other's minds is called *Theory of Mind*, inferring what others are thinking or feeling. *Theory of Mind* answers the question, *What is the other person thinking and feeling?* which actually implies that you have answered a more fundamental question, *Does the other person think and feel?* How we answer this second question is referred to as mind perception, or mind attribution, and forms an important foundation of human–AI comparison.

Mind perception is often divided into capacities, not surprisingly, related to thinking and feeling. Related to thinking is an *agency*, the capacity to act, think, plan, decide, reason, and have intuition and goals. Related to feeling is an *experience*, the capacity to feel emotions, experience pain and pleasure, and have memories, desires, and beliefs (Gray et al., 2007; Wegner & Gray, 2017).

We assume a complete mind—maximum levels of both agency and experience—when observing, comparing, or interacting with neurotypical adult humans. We care about what actors are feeling because we know they can feel a full range of emotions and we care what criminals planned because we know they have the capacity for premeditated crimes. In contrast, when a *Saturday Night Live* sketch includes animals in characteristically human roles it's funny precisely because the animals can't understand complex social situations and feel complex emotions in response. Likewise, rulings in court are quite different for crimes by minors or by adults judged *not guilty by reason of insanity* compared to fully neurotypical minded adults.

Therefore, mind perception is important for differentiating between different types of entities such as nonhumans like animals and machines and different classes of humans like children and adults (Wegner & Gray, 2017). We tend to see children, babies, and complex animals as having very high—even maximum—capacities for experience similar to human adults, but unlike adults, we perceive them as having less agency. We perceive simpler less humanlike animals—like vermin and bugs—as having much less experience as well as agency. Interestingly, dead humans, those in long-term comas, and those with different neurological disorders tend to also have intermediate levels of agency and experience, differing on a case-to-case basis. We generally perceive God and corporations as having high agency, but nearly no experiential mind. Finally, robots and AI are perceived as having a moderate level of agency and hardly any experience capacity (Gray et al., 2007; Wegner & Gray, 2017), making them different from humans in both dimensions of mind perception.

Not only do we believe AI collectively have different minds from humans, but specific situations can alter the perceived mind of an individual.

First, while humans are high in both agency and experience, we tend to discount one if the other is emphasized. Someone who has powerfully changed the world—high agency—is seen as being less susceptible to pain—lower experience. People rated Osama bin Laden, Adolf Hitler, Mother Teresa, and Martin Luther King Jr. as high in agency, but as also being less susceptible to pain. In contrast, someone who's suffered and been victimized—high experience—is thought to be held less responsible—lower agency. People rated orphans, victims of date rape, and a person with Down's Syndrome as being more susceptible to pain, but held less blameworthy or praiseworthy for their actions (Gray & Wegner, 2009).

Second, as hinted in this last example, this mind attribution process is related to morality. Moral agents, like perpetrators, need at least some agency and moral victims need at least some capacity for experience or we don't consider it a moral act (Gray et al., 2012). Likewise, being a perpetrator or victim in a moral act increases one's agency or experience, respectively (Gray et al., 2014; Gray et al., 2012).

Third, situations unrelated to morality can change our overall perceptions of other's minds. For example, we perceive communicating with a spoken voice as indicating more mind compared to the same communication with just text (Schroeder et al., 2017). Similarly, we perceive more mind in job candidates interviewing face-to-face compared to those interviewing over video (Baker et al., 2020).

While we typically perceive machines as moderate in agency and low in experience, specific machines and situations related to appearance, identity, and behavior can alter that. When robots or characters have humanlike faces (Appel et al., 2020; Gray & Wegner, 2012; Martini et al., 2016; Zhao et al., 2019) or voices (Schroeder & Epley, 2016), a name associated with one's ingroup (Eyssel & Kuchenbrandt, 2012), the ability to experience emotion (Bigman & Gray, 2018; Gray & Wegner, 2012), when we help them (Tanibe et al., 2017), or when they are harmed

(Ward et al., 2013; Wieringa et al., 2024), we perceive them as having a mind capable of greater experiences. When machines are anthropomorphized with faces (Martini et al., 2016), voices, names, and gender (Waytz et al., 2014), described as independent (Gray & Wegner, 2012), and when they act unpredictably (Waytz, Morewedge, et al., Waytz, Morewedge, et al., 2010), or commit moral wrongs (Shank & DeSanti, 2018), we perceive them as having greater agency.

However, these situations rarely cause us to perceive a machine as a fully minded human, which suggests that we perceive some immutable differences in the mind of humans compared to machines. Comparisons involving these unchanging differences in beliefs make up the essence of being a human or an AI.

4.6 Essence

The essence or the nature of something is what it is at its core, the fundamental, immutable properties that define it. Psychological essentialism is what people believe about the essence or nature of different types of entities. It includes beliefs like humans have souls, and that humans having souls differentiates us from nonhuman animals and machines.

Therefore, most people tend to find that AI along with other machines have a different essential nature than humans. This is often rooted in differences in being "biological" versus "mechanical" (Haslam et al., 2000) and this is almost entirely about long-term beliefs, and little about what we observe in specific situations (Fig. 4.1).

Essence relates to important characteristics that we ascribe to being fundamental to human identity, associated with biological (Haslam, 1998) and personality traits (Haslam et al., 2004). Thus, people's essentialist beliefs are used to stereotype (Bastian & Haslam, 2006) and devalue (Haslam et al., 2000) other humans, specifically social outgroups (Haslam et al., 2005; Leyens et al., 2001). At its extreme, people fully deny essential human characteristics to other humans, a phenomenon called dehumanization (Haslam, Loughnan, et al., Haslam, Loughnan, et al., 2008), which in many ways is the opposite of anthropomorphism.

Dehumanization can happen in two different ways, which is important for understanding human–AI differences. Denying unique human characteristics related to culture, morality, rationality, and civility leads to animalistic dehumanization, while denying human nature characteristics related to individuality, interpersonal warmth, and mental creativity leads to mechanistic dehumanization (Haslam, 2006; Haslam, Loughnan, et al., Haslam, Loughnan, et al., 2008), interestingly named after machines.

These essential attributes of human nature like being friendly, jealous, aggressive, fun-loving, or trusting divide humans and animals on one side from machines and robots on the other (Loughnan & Haslam, 2007). Many of these essential human–machine differences relate to emotions and aspects of experience, and these differences persist across many human cultures (Haslam, Kashima, et al., Haslam, Kashima, et al., 2008). Therefore, while different psychological concepts, mind and essence often functionally point us to the same human–AI comparison.

As perceivers, our comparison of human–AI differences in particular situations often refer to these essential differences. In a study Jenny Davis and I did on role-taking, participants read an emotional first-person narrative attributed to an AI or human. Then we asked them a simple question of what the narrator was thinking and feeling when sharing the story (Davis et al., 2022).

Using essentialist language, some participants categorically denied the essential ability for AI to think and feel, writing, "I do not think that Rob was thinking and feeling, because he is an AI robot" and "Rob is a robot so in my opinion 'he' was not thinking or feeling anything. I guess it boils down to whether or not I think highly advanced AI robots can think or feel. I don't think so. The 'thoughts' and 'feelings' of a robot are programming constructs." (p. 14).

Some other participants also denied this essence through an explicit comparison to the capabilities of a human, "If Rebecca were a human being, I believe that she would be feeling overwhelmed and stressed by the conditions of her workplace. I believe that if she were not an AI she would be thinking of finding a different job" (Davis et al., 2022; p. 15).

These explanations showcase that, even with differences, people generally agreed in their essential beliefs about AI as not having full capacity to think and feel (Pauketat & Anthis, 2022). But this does not mean that our collective essentialist beliefs about AI cannot change with cultural and technological development.

A team of researchers at the Sentience Institute have been asking critical questions about the essence of AI, including in 2021 and 2023. In 2023, they found that 19% of people thought that current AI was sentient while 38% agreed that AI could one day be sentient, and this was up from 18% and 35%, respectively, from 2021. Likewise, perceptions of AI's ability to think analytically, part of agency, went from a 62.1 rating out of 100 in 2021 to 67.1 in 2023, while experiencing emotions, part of experiential mind, went from 34.3 to 36.8 over the same period (Pauketat et al., 2023). Among other developments, ChatGPT was released in this two-year period, so collective shifts in AI essence perceptions, while slow, are not impossible or imperceptible.

Even with individual and societal shifts over time, essential human–AI comparisons are based heavily on our beliefs (Fig 4.1) and, as such, have an undue and more stable influence as a cause of the machine penalty. Other causes like appearance and identity are more likely to influence us implicitly—like the mindlessness mechanism suggested by *Computers Are Social Actors*—so that our immediate, socially expected reaction is based on the cues we receive.

However, essentialist beliefs—and mind perception to some degree—affect human–AI comparisons through a more deliberate thinking process, such as when people are asked to give an explanation about an AI. Like selecting a piece of cake, we may go with your gut instinct based on the appearance of the cake. In contrast, we may deliberate, observe, consider, and even seek more information. Regardless of whether one or all of these comparisons come into play, the result is differentiation of machines from humans. And this leads us to explore how we diminish machines.

References

Appel, M., Izydorczyk, D., Weber, S., Mara, M., & Lischetzke, T. (2020). The uncanny of mind in a machine: Humanoid robots as tools, agents, and experiencers. *Computers in Human Behavior, 102*, 274–286. https://doi.org/10.1016/j.chb.2019.07.031

Baker, D. A., Burns, D. M., & Reynolds Kueny, C. (2020). Just Sit Back and Watch: Large Disparities between Video and Face-to-face Interview Observers in Applicant Ratings. *International Journal of Human–Computer Interaction, 36*(20), 1968–1979. https://doi.org/10.1080/10447318.2020.1805874

Bastian, B., & Haslam, N. (2006). Psychological essentialism and stereotype endorsement. *Journal of Experimental Social Psychology, 42*(2), 228–235. https://doi.org/10.1016/j.jesp.2005.03.003

Bigman, Y. E., & Gray, K. (2018). People are averse to machines making moral decisions. *Cognition, 181*, 21–34. https://doi.org/10.1016/j.cognition.2018.08.003

Burke, P. J., & Stets, J. E. (2009). *Identity theory*. Oxford University.

Carr, E. W., Hofree, G., Sheldon, K., Saygin, A. P., & Winkielman, P. (2017). Is that a human? Categorization (dis)fluency drives evaluations of agents ambiguous on human-likeness. *Journal of Experimental Psychology: Human Perception and Performance, 43*(4), 651–666. https://doi.org/10.1037/xhp0000304

Caruana, R., Lou, Y., Gehrke, J., Koch, P., Sturm, M., & Elhadad, N. (2015). Intelligible models for healthcare: Predicting pneumonia risk and hospital 30-day readmission. *Proceedings of the 21th ACM SIGKDD international conference on knowledge discovery and data mining*, 1721–1730. https://doi.org/10.1145/2783258.2788613

Christian, B. (2011). *The most human human: What TALKING WITH COMPUTERS TEACHES US ABOUT WHAT IT MEANS TO BE ALIVE*. Anchor Books.

Davis, J. L., Shank, D. B., Love, T. P., Stefanik, C., & Wilson, A. (2022). Gender dynamics in human-AI role-taking. In W. Kalkhoff, S. R. Thye, & E. J. Lawler (Eds.), *Advances in group processes* (Vol. 39, pp. 1–22). Emerald Publishing Limited. https://doi.org/10.1108/S0882-614520220000039001

Epley, N. (2015). *Mindwise: Why we misunderstand what others think, believe, feel, and want*. Vintage.

Epstein, R., Roberts, G., & Beber, G. (Eds.). (2008). *Parsing the turing test: Philosophical and methodological issues in the quest for the thinking computer*. Springer.

Eyssel, F., & Kuchenbrandt, D. (2012). Social categorization of social robots: Anthropomorphism as a function of robot group membership. *British Journal of Social Psychology, 51*(4), 724–731. https://doi.org/10.1111/j.2044-8309.2011.02082.x

Feine, J., Gnewuch, U., Morana, S., & Maedche, A. (2020). Gender bias in Chatbot design. *Chatbot Research and Design, 11970*, 79–93. https://doi.org/10.1007/978-3-030-39540-7_6

Fortunati, L., Edwards, A., Edwards, C., Manganelli, A. M., & de Luca, F. (2022). Is Alexa female, male, or neutral? A cross-national and cross-gender comparison of perceptions of Alexa's gender and status as a communicator. *Computers in Human Behavior, 137*, 107426. https://doi.org/10.1016/j.chb.2022.107426

Gray, H. M., Gray, K., & Wegner, D. M. (2007). Dimensions of mind perception. *Science, 315*(5812), 619. https://doi.org/10.1126/science.1134475

Gray, K., Schein, C., & Ward, A. F. (2014). The myth of harmless wrongs in moral cognition: Automatic dyadic completion from sin to suffering. *Journal of Experimental Psychology: General, 143*(4), 1600–1615. https://doi.org/10.1037/a0036149

Gray, K., & Wegner, D. M. (2009). Moral typecasting: Divergent perceptions of moral agents and moral patients. *Journal of Personality and Social Psychology, 96*(3), 505–520. https://doi.org/10.1037/a0013748

Gray, K., & Wegner, D. M. (2012). Feeling robots and human zombies: Mind perception and the uncanny valley. *Cognition, 125*(1), 125–130. https://doi.org/10.1016/j.cognition.2012.06.007

Gray, K., Young, L., & Waytz, A. (2012). Mind perception is the essence of morality. *Psychological Inquiry, 23*(2), 101–124. https://doi.org/10.1080/1047840X.2012.651387

Haslam, N. (1998). Natural kinds, human kinds, and essentialism. *Social Research, 65*(2), 291–314.

Haslam, N. (2006). Dehumanization: An integrative review. *Personality and Social Psychology Review, 10*(3), 252–264. https://doi.org/10.1207/s15327957pspr1003_4

Haslam, N., Bain, P., Douge, L., Lee, M., & Bastian, B. (2005). More human than you: Attributing humanness to self and others. *Journal of Personality and Social Psychology, 89*(6), 937–950. https://doi.org/10.1037/0022-3514.89.6.937

Haslam, N., Bastian, B., & Bissett, M. (2004). Essentialist beliefs about personality and their implications. *Personality and Social Psychology Bulletin, 30*(12), 1661–1673. https://doi.org/10.1177/0146167204271182

Haslam, N., Kashima, Y., Loughnan, S., Shi, J., & Suitner, C. (2008). Subhuman, inhuman, and superhuman: Contrasting humans with nonhumans in three cultures. *Social Cognition, 26*(2), 248–258. https://doi.org/10.1521/soco.2008.26.2.248

Haslam, N., Loughnan, S., Kashima, Y., & Bain, P. (2008). Attributing and denying humanness to others. *European Review of Social Psychology, 19*(1), 55–85. https://doi.org/10.1080/10463280801981645

Haslam, N., Rothschild, L., & Ernst, D. (2000). Essentialist beliefs about social categories. *British Journal of Social Psychology, 39*(1), 113–127. https://doi.org/10.1348/014466600164363

Heise, D. R. (2007). *Expressive order: Confirming sentiments in social actions.* Springer.

Kätsyri, J., Förger, K., Mäkäräinen, M., & Takala, T. (2015). A review of empirical evidence on different uncanny valley hypotheses: Support for perceptual mismatch as one road to the valley of eeriness. *Frontiers in Psychology, 6*, 390. https://doi.org/10.3389/fpsyg.2015.00390

Khanna, N. (2010). Passing as black: Racial identity work among biracial Americans. *Social Psychology Quarterly, 73*(4), 380–397. https://doi.org/10.1177/0190272510389014

Kim, B., de Visser, E., & Phillips, E. (2022). Two uncanny valleys: Re-evaluating the uncanny valley across the full spectrum of real-world human-like robots. *Computers in Human Behavior, 135*, 107340. https://doi.org/10.1016/j.chb.2022.107340

Kinzler, K. D. (2021). Language as a social cue. *Annual Review of Psychology, 72*, 241–264. https://doi.org/10.1146/annurev-psych-010418-103034

Leyens, J. P., Rodriguez-Perez, A., Rodriguez-Torres, R., Gaunt, R., Paladino, M. P., Vaes, J., & Demoulin, S. (2001). Psychological essentialism and the differential attribution of uniquely human emotions to ingroups and outgroups. *European Journal of Social Psychology, 31*(4), 395–411. https://doi.org/10.1002/ejsp.50

Löffler, D., Dörrenbächer, J., & Hassenzahl, M. (2020). The uncanny valley effect in zoomorphic robots: The U-shaped relation between animal likeness and likeability. In *Proceedings of the 2020 ACM/IEEE international conference on human-robot interaction* (pp. 261–270). https://doi.org/10.1145/3319502.3374788

Loughnan, S., & Haslam, N. (2007). Animals and androids: Implicit associations between social categories and nonhumans. *Psychological Science, 18*(2), 116–121. https://doi.org/10.1111/j.1467-9280.2007.01858.x

MacDorman, K. F. (2024). Does mind perception explain the uncanny valley? A meta-regression analysis and (de)humanization experiment. *Computers in Human Behavior: Artificial Humans, 2*(1), 100065. https://doi.org/10.1016/j.chbah.2024.100065

MacKinnon, N. J., & Heise, D. R. (2010). *Self, identity, and social institutions*. Palgrave Macmillan.

Malle, B. F., Magar, S. T., & Scheutz, M. (2019). AI in the sky: How people morally evaluate human and machine decisions in a lethal strike dilemma. In I. A. Ferreira, J. S. Sequeira, G. S. Virk, E. E. Kadar, & O. Tokhi (Eds.), *Robots and well-being* (pp. 111–133). Springer. https://doi.org/10.1007/978-3-030-12524-0_11

Manoli, A., Pauketat, J. V. T., & Anthis, J. R. (2025). The AI double standard: Humans judge all AIs for the actions of one. Conference on computer-supported cooperative work, Yokohama, Japan.

Martini, M. C., Gonzalez, C. A., & Wiese, E. (2016). Seeing minds in others – Can agents with robotic appearance have human-like preferences? *PloS ONE, 11*(1), e0146310. https://doi.org/10.1371/journal.pone.0146310

Mathur, M. B., & Reichling, D. B. (2016). Navigating a social world with robot partners: A quantitative cartography of the Uncanny Valley. *Cognition, 146*, 22–32. https://doi.org/10.1016/j.cognition.2015.09.008

Mathur, M. B., Reichling, D. B., Lunardini, F., Geminiani, A., Antonietti, A., Ruijten, P. A., et al. (2020). Uncanny but not confusing: Multisite study of perceptual category confusion in the Uncanny Valley. *Computers in Human Behavior, 103*, 21–30. https://doi.org/10.1016/j.chb.2019.08.029

McCoy, J. P., & Ullman, T. D. (2018). A minimal turing test. *Journal of Experimental Social Psychology, 79*, 1–8. https://doi.org/10.1016/j.jesp.2018.05.007

Metz, C. (2016). Google's AI wins pivotal second game in match with go grandmaster. *Wired Magazine*. https://www.wired.com/2016/03/googles-ai-wins-pivotal-game-two-match-go-grandmaster/

Mori, M. (1970). The uncanny valley. *Energy, 7*(4), 33–35.

Nass, C., Reeves, B., & Leshner, G. (1996). Technology and roles: A tale of two TVs. *Journal of Communication, 46*(2), 121–128. https://doi.org/10.1111/j.1460-2466.1996.tb01477.x

Neufeld, E., & Sonje, F. (2020). In defense of the Turing test. *AI & Society, 35*(4), 819–827. https://doi.org/10.1007/s00146-020-00946-8

Owens, T. J., Robinson, D. T., & Smith-Lovin, L. (2010). Three faces of identity. *Annual Review of Sociology, 36*, 477–499. https://doi.org/10.1146/annurev.soc.34.040507.134725

Palomäki, J., Kunnari, A., Drosinou, M., Koverola, M., Lehtonen, N., Halonen, J., et al. (2018). Evaluating the replicability of the uncanny valley effect. *Heliyon, 4*(11), e00939. https://doi.org/10.1016/j.heliyon.2018.e00939

Pauketat, J. V. T., & Anthis, J. R. (2022). Predicting the moral consideration of artificial intelligences. *Computers in Human Behavior, 136*, 107372. https://doi.org/10.1016/j.chb.2022.107372

Pauketat, J. V. T., Ladak, A., & Anthis, J. R. (2023). Artificial Intelligence, Morality, and Sentience (AIMS) survey: 2023 update. *Center for Open Science*. https://doi.org/10.31234/osf.io/9xsav

Phillips, E., Zhao, X., Ullman, D., & Malle, B. F. (2018). What is human-like? Decomposing robots' human-like appearance using the anthropomorphic roBOT (ABOT) database. In *Proceedings of the 2018 ACM/IEEE international conference on human-robot interaction* (pp. 105–113). https://doi.org/10.1145/3171221.3171268

Schroeder, J., & Epley, N. (2016). Mistaking minds and machines: How speech affects dehumanization and anthropomorphism. *Journal of Experimental Psychology: General, 145*(11), 1427–1437. https://doi.org/10.1037/xge0000214

Schroeder, J., Kardas, M., & Epley, N. (2017). The humanizing voice: Speech reveals, and text conceals, a more thoughtful mind in the midst of disagreement. *Psychological Science, 28*(12), 1745–1762. https://doi.org/10.1177/0956797617713798

Schwind, V., Leicht, K., Jäger, S., Wolf, K., & Henze, N. (2018). Is there an uncanny valley of virtual animals? A quantitative and qualitative investigation. *International Journal of Human-Computer Studies, 111*, 49–61. https://doi.org/10.1016/j.ijhcs.2017.11.003

Shank, D. B., & DeSanti, A. (2018). Attributions of morality and mind to artificial intelligence after real-world moral violations. *Computers in Human Behavior, 86*, 401–411. https://doi.org/10.1016/j.chb.2018.05.014

Shank, D. B., Graves, C., Gott, A., Gamez, P., & Rodriguez, S. (2019). Feeling our way to machine minds: People's emotions when perceiving mind in artificial intelligence. *Computers in Human Behavior, 98*, 256–266. https://doi.org/10.1016/j.chb.2019.04.001

Song, Y., Luximon, A., & Luximon, Y. (2021). The effect of facial features on facial anthropomorphic trustworthiness in social robots. *Applied Ergonomics, 94*, 103420. https://doi.org/10.1016/j.apergo.2021.103420

Stein, J.-P., & Ohler, P. (2017). Venturing into the uncanny valley of mind – The influence of mind attribution on the acceptance of human-like characters in a virtual reality setting. *Cognition, 160*, 43–50. https://doi.org/10.1016/j.cognition.2016.12.010

Tanibe, T., Hashimoto, T., & Karasawa, K. (2017). We perceive a mind in a robot when we help it. *PloS One, 12*(7), e0180952. https://doi.org/10.1371/journal.pone.0180952

Thepsoonthorn, C., Ogawa, K.-i., & Miyake, Y. (2021). The exploration of the uncanny valley from the viewpoint of the robot's nonverbal behaviour. *International Journal of Social Robotics, 13*, 1443–1455. https://doi.org/10.1007/s12369-020-00726-w

Tinwell, A. (2014). *The uncanny valley in games and animation*. CRC Press.

Turing, A. M. (1950). Computing, machinery, and intelligence. *Mind, 59*(236), 433–460.

Wang, S., Lilienfeld, S. O., & Rochat, P. (2015). The uncanny valley: Existence and explanations. *Review of General Psychology, 19*(4), 393–407. https://doi.org/10.1037/gpr0000056

Ward, A. F., Olsen, A. S., & Wegner, D. M. (2013). The harm-made mind: Observing victimization augments attribution of minds to vegetative patients, robots, and the dead. *Psychological Science, 24*(8), 1437–1445. https://doi.org/10.1177/0956797612472343

Waytz, A., Heafner, J., & Epley, N. (2014). The mind in the machine: Anthropomorphism increases trust in an autonomous vehicle. *Journal of Experimental Social Psychology, 52*, 113–117. https://doi.org/10.1016/j.jesp.2014.01.005

Waytz, A., Morewedge, C. K., Epley, N., Monteleone, G., Gao, J.-H., & Cacioppo, J. T. (2010). Making sense by making sentient: Effectance motivation increases anthropomorphism. *Journal of Personality and Social Psychology, 99*(3), 410–435. https://doi.org/10.1037/a0020240

Wegner, D. M., & Gray, K. (2017). *The mind club: Who thinks, what feels, and why it matters*. Penguin.

Wieringa, M. S., Müller, B. C. N., Bijlstra, G., & Bosse, T. (2024). Robots are both anthropomorphized and dehumanized when harmed intentionally. *Communications Psychology, 2*(1), 72. https://doi.org/10.1038/s44271-024-00116-2

Zhao, X., & Malle, B. F. (2022). Spontaneous perspective taking toward robots: The unique impact of humanlike appearance. *Cognition, 224*, 105076. https://doi.org/10.1016/j.cognition.2022.105076

Zhao, X., Phillips, E., & Malle, B. F. (2019). *How people infer a humanlike mind from a robot body*. https://psyarxiv.com/w6r24/download

5

How Do We Diminish Machines?

Whether you look around at occurrences in everyday life or you look at the research literature, there are plenty of examples demonstrating the machine penalty, that *comparing AI to humans leads us to diminish similar outcomes from AI across situations.* In this chapter, I'll track the history of diminishing machines from hints from the past to empirical studies from the modern research literature, all pointing to our tendency to diminish machines. Perhaps more importantly I'll begin to consider not just that we do tend to diminish machines, but how and when we do, which will lead to the machine penalty causes in the next chapter and applications in Part II.

5.1 Diminishing Computation Throughout History

A broad view of history indicates that diminishing machines and computational output is not a phenomenon dependent on AI, but on computation, calculation, and quantification more generally. People rejected conclusions based on quantitative observation by scientists like Galileo,

D. B. Shank, *The Machine Penalty*, https://doi.org/10.1007/978-3-031-86061-4_5

Copernicus, Newton, and Darwin for the more normative beliefs of the time. Likewise, germ theory, quantum physics, and chemical and molecular reactions all rely on calculations that the average person might not understand, and their conclusions often seemed at odds with general human intuition.

The infamous Luddite rebellion of the early 1800s was another case of humans rejecting machines, in this case the loom machines that were taking their jobs. The protests and clandestine raids against machine-based automation were not a simple rejection of machines, but a deadly fight for workers' rights against powerful factory owners (Andrews, 2023 [2015]; Conniff, 2011). Therefore, today the term *luddite* refers more generally to those who fight against automation or reject technology more broadly. While neither scientific theories that opposed the religious worldview of their day nor rebellion against job loss due to automation are the machine penalty proper, a bird's eye view of history still shows antagonism, disbelief, and a rejection of disagreeable outcomes rooted in computation and machines.

5.1.1 The Late 1700s' Puppet-Doll Chess Scam

One of the earliest examples of diminishing an AI feels more like clickbait than actual events from the late 1700s and early 1800s. The headline could read, "Giant Chess-Playing Doll Scams Benjamin Franklin, Napoleon Bonaparte; Foiled by Edgar Allan Poe."

In 1770, a Hungarian inventor named Wolfgang von Kempelen created an automaton—what we might call a life-size mechanical puppet-doll—that he claimed was the first thinking machine as it could play chess. Known as the *Mechanical Turk*, it had a humanlike body and face, was dressed in ornate silk robes and a turban, and was connected to one side of a table with cabinets that housed its mechanics below and the chess board on top. People could play it in chess, and it would usually win (Brahambhatt, 2022; Morton, 2015; Stephens, 2023).

As it did, its popularity and the mystery around it grew, so Kempelen and his successor took it on tours of Europe. Among others, it played and defeated all the big names of the day—Napoleon Bonaparte, Benjamin

Franklin, and Charles Babbage, creator of one of the first computers ever. The primary reaction of those who played and observed it was disbelief and a search for alternative explanations.

They inherently didn't think a machine could play chess that well. But on seeing the outcomes of superior play along with its correcting human opponent's illegal moves—and foiling Napoleon's attempts at cheating—observers and players found other ways to disparage it. They accused Kempelen of controlling it from the side using magnets or strings, or that a human—perhaps a child—was in the cabinet running the mechanical controls, although opening it revealed no such person. While the *Mechanical Turk* was *actually* a hoax, people's reactions to it were real: a strong sense of believing that machines should be less than humans.

While speculation about it being a hoax continued for many years, for some time no one was able to find evidence or present a convincing alternative explanation. But eventually someone did, and that whistleblower was none other than Edgar Allan Poe, the writer famous for *The Raven* and *The Tell-Tale Heart.* He penned an essay explaining the *Mechanical Turk's* mechanisms, how a small human could have controlled it from inside a secret chamber, where that chamber was, and why the *Mechanical Turk's* behavior was more in line with a human than a machine. While the *Mechanical Turk* was never definitely proven to be a hoax, that effectively ended popular interest in the device (Brahambhatt, 2022; Morton, 2015; Stephens, 2023).

5.1.2 Diminishing in Recent History

Interestingly, when computers could actually rival humans in chess, the disbelief and search for alternatives sometimes persisted. When, in 1997, world chess champion Garry Kasparov was defeated by Deep Blue, IBM's AI chess program, he responded with skepticism about Deep Blue's abilities, suggesting that IBM had created a machine to beat him specifically and questioning whether all the moves originated with Deep Blue (Chandrasekaren, 1997).

As we moved into the period of the modern computer, there has been increased potential for the diminishing of machines and their outputs, as computations became more complex and more hidden from view from their end users. One systematic literature review of people's acceptance of computational versus human advice considered research articles extending as far back as 1950 (Burton et al., 2020), near the time of the creation of the first modern computers.

From those early days of computer science, algorithms were designed to make predictions or decisions that humans would otherwise make, especially ones that were difficult for humans. Many of the reasons computers were developed were to make predictions about missile trajectories, spatial bodies, or weather systems. And from those early days, people diminished, were skeptical of, and rejected those outputs in spite of these algorithms being created to solve difficult problems and doing so better than humans (Dawes, 1979; Meehl, 1954).

5.2 My Uncovering the Penalty

5.2.1 First Evidence

I first encountered evidence that we diminish machines during my graduate research in social psychology at the University of Georgia. In my first experiment, university student participants engaged in an "online" exchange game over a computer network with other players (Shank, 2008). Over a series of rounds, each player had the option to either reward or punish only one of two other players.

There were two experimental conditions related to the behavior of one of the other players. First, that player was either labeled as a computer or another student. Second, this other player used a baseline cooperative strategy that was mainly *tit-for-tat*—reward in response to reward; punishment for punishment; nonexchange for nonexchange—or a coercive strategy that involved high levels of punishment any time the participant failed to reward them.

Participants perceived the baseline cooperative behavior as no different in fairness for both human and computer players. That is, there was no penalty. As expected, they saw coercive behavior as unfair. But participants diminished the level of unfairness when rating the computer player's coercive behavior as less unfair than the human player's coercive behavior (Shank, 2012b).

This difference may have been, in part, because cooperative behavior was desirable and expected in this situation, whereas coercive behavior was detrimental and unexpected. People pay more attention to negative outcomes (Baumeister et al., 2001) and their causes (Morewedge, 2009), suggesting that participants would care more about coercive condition outcomes, and thereby distinguish more by who or what caused it.

Likewise, participants' *behavior*—as opposed to *perception*—supported the machine penalty for coercive behavior, while also giving me the first glimpse of the consequences. I measured participants' exchange behaviors including resistance—which was withholding rewards from their partner—and punishment. Resistance and punishment were higher toward coercive agents compared to cooperative agents, but the levels again differed by agent's identity. For both behaviors, participants resisted and punished coercive humans more than coercive computers (Shank, 2012b).

5.2.2 Documenting the Penalty for Positive and Negative Outcomes

I wanted to examine this more thoroughly including the implications, so I set out to do another experiment that could confirm these general findings and extend them. This experiment involved a cover story that I was testing an e-commerce platform—like eBay or Amazon—and that the participant would interact with either a human or computer representative of a real company. Over each round, in a series of rounds, student participants would get to decide how much of a generic "product" they would buy from the representative. Then they "received" the products by finding out their quality, before proceeding to the next round.

In experiments, there is always a concern that participants won't pay attention, take it seriously, or be as motivated as they would in real life. To combat this, I paid participants extra based on the quality of products they bought. If the products were high-quality, they'd earn more actual money by buying more of the product, but if they were low-quality, they would earn more actual money by buying less of the product. I experimentally manipulated the identity of the representative as a human or computer, the quality of the product as low or high, and the knowledge of the representative's relationship to the company. I measured outcomes such as emotions, impressions, and future interaction (Shank, 2012a).

Not unexpectedly, people reported more positive emotions and impressions and less negative emotions when receiving high-quality products compared to low-quality ones. No one likes to find the new sweater, cat feeder, or organizer they ordered doesn't live up to the quality they were expecting. On average, it didn't make a difference if the products were being provided by a computer agent or a human employee.

But the averages obscured the really interesting finding: some of the emotions, like being upset at the computer representatives, were less variable than the emotions felt toward human representatives—less variable in two directions: in both feeling upset and not feeling upset. Participants were moderately upset toward the representative after receiving low-quality products. This was about 60% on the upset scale when the agent was a human, but only about 50% on that scale when the agent was a computer. When receiving high-quality products, participants weren't very upset and scored low on that scale. But they scored lower for humans at about 15%, whereas they were 25% upset for high-quality products from computer representatives (Shank, 2013). Therefore, participants expressed a greater range toward humans (15% to 60%) than computer agents (25% to 50%). The extreme values of being upset or not-upset were diminished when the behavior came from a machine.

In line with the *Relevant Outcomes Condition*, the outcomes need to be relevant to a situation, and opposite behavior led to relevant outcomes that were opposites. I expected that low-quality products would lead people to be somewhat upset, and that would be diminished when they came from computer agents. That's exactly what I found. What I didn't anticipate, but is more obvious in hindsight, is that high-quality products should lead to being more "not-upset"—that's the relevant outcome—and that computer agents should diminish how "not upset" one feels. And they did, with human's closer to being not upset (15%) than computers (25%). So, the machine penalty can work on completely opposite outcomes if those are each relevant in opposite situations.

Also, this result and concept help explain the finding in my first experiment of no difference in the fairness of cooperative behavior. In that set up, cooperative behavior was a sort of neutral, default behavior and even expected: for example, the other player the participant could interact with always had a cooperative strategy. That made it seem normative. Coercive behavior was the negative outcome, whereas cooperative behavior was a near neutral outcome, and if I would have included generous behavior, it would have been a truly positive outcome. So, there wasn't much of a relevant outcome to diminish. Other research in the purchasing domain finds parallel results, with better-than-expected offers from machines accepted less often when from a machine, but worse-than-expected offers accepted more often when from a machine (Garvey et al., 2023).

For the second experiment, the results I've discussed were for the control condition where no information was given for the relationship between the human or computer agent and the company. In other conditions, I informed participants either that the representative was restricted and regulated by the company or that they were independent from the company. When this information was included, a different application of the penalty was observed.

As expected, agents that were restricted by their company were perceived as less powerful, responsible, and in control, compared to those who were more independent from their company. But independent computers were less powerful, responsible, and in control (54%–59%) than independent humans (63%–68%). Conversely, regulated computers were not as low in power, responsibility, and control (48%–57%) compared to regulated humans, which were less powerful when regulated (30%–31%) (Shank, 2014). Again, outcomes from machines are diminished compared to their human counterparts, and the outcomes can be opposite outcomes, based on their relevance to that situation.

I also have research beyond this early work that demonstrates the machine penalty—and most of that is covered somewhere in this book. However, I wasn't the only researcher who decided to compare machines to people. Several others also uncovered the machine penalty, and collectively this led to a new research area.

5.3 Algorithm Aversion and Appreciation

In the last decade research comparing AI to humans has ballooned (Mahmud et al., 2022). Some of this is also due to new terms for it, such as *algorithm aversion* coined in a research article by Dietvorst, Simmons, and Massey first published online in 2014 entitled, *Algorithm Aversion: People Erroneously Avoid Algorithms After Seeing Them Err.*

Their provocative idea supported in five studies was that when people see algorithms—what we might just as easily call AI or machines—make incorrect predictions, those people choose them less often to make future predictions, compared to humans whom they've seen make even worse predictions. In their experiments, participants either forecasted the success of MBA applicants or predicted the ranking of states in terms of airline passengers. Participants earned more money for better predictions, but in spite of that, they preferred imperfect humans over imperfect machines, even though the imperfect machine predictions were clearly superior (Dietvorst et al., 2015).

Often positioned as the opposite of algorithm aversion, *algorithm appreciation* was another new term associated with the argument that people prefer and value AI more than humans. This term was introduced by Jennifer Logg and her colleagues who conducted six experiments comparing advice from humans versus algorithms (Logg et al., 2019).

They found across different types of tasks that more people relied on the algorithm's advice over advice from another person or group of people, including on some tasks that were more subjective and related to relationships or aesthetics. Their primary outcome was influence—how much participants shifted their own position toward that of the algorithm or human. Interestingly, the preference for algorithms was reduced when the decision was made by experts or when the participant's own prediction was the alternative to the algorithm (Logg et al., 2019), suggesting that comparisons of algorithm's predictions are strongly affected by what human they are pitted against (*AI vs Human Condition*).

While the terms *algorithm aversion* and *algorithm appreciation* are used in general as opposites, the results of these original studies were actually quite compatible. Logg and colleagues point that Dietvorst and colleagues' original algorithm aversion study that the effect was only after performance information, and the comparison was usually with oneself, not to a random human. Critically, only in Dietvorst and colleagues' study 4 did they compare the participant's decision between the AI algorithm and a human that wasn't oneself, in this case a subject who had previously done their experiment. In this case, when no information about the algorithm's or human's performance was shown, about two-thirds of participants favored the algorithm (Dietvorst et al., 2015), supporting algorithmic appreciation.

Yet when both the human's and algorithm's prior performances were shown, participants were virtually split with half selecting the human and half selecting the algorithm. Now this tie might seem like it points to no effect—neither aversion nor appreciation—but it's actually evidence for algorithm aversion for two reasons. First, this was a drop from the baseline which showed two-thirds of participants selected the algorithm. Second, the performance of the algorithm was on average better than the human, so there was an additional rational reason to select it instead of the human. Therefore, in just this one study, we can see how algorithm aversion and appreciation—what I call the machine penalty and advantage—can both be supported, under slightly different conditions. So, which one is right?

5.4 Diminishing Machines Today

There are now hundreds of research studies that consider people's reactions to AI versus human prediction, advice, creations, and other outputs and, therefore, provide evidence for the machine penalty, the machine advantage, neither, or possibly sometimes both. Some of these studies specifically label this as algorithm aversion or algorithm appreciation whereas others adopt different terms (Mahmud et al., 2022). Fortunately, three systematic literature reviews (Burton et al., 2020; Jussupow et al., 2020; Mahmud et al., 2022)—research papers that synthesize the empirical research literature in an area—and one meta-analysis (Qin et al., in press)—a research paper that statistically combined the results from the empirical research literature—help show when people diminish machines.

From that literature, we might want to ask and try to answer the questions: *Are machines' outcomes generally perceived better or worse? Is the machine penalty or the machine advantage actually correct?* Or at least, *Which has the most support?* Across the research in this area, there is more empirical support and more cases that show people diminish machines compared to humans (Burton et al., 2020; Jussupow et al., 2020; Mahmud et al., 2022; Qin et al., in press)—thus the reason this book is named *The Machine Penalty*.

However, those questions undermine what really matters and that is *when* one or the other is supported or what conditions lead to one being the stronger, more dominant force. It's analogous to the oft-debated question, *does social media have a positive or negative effect on wellbeing?* The answer to that is: *Both, depending on how it's used and what's measured* (Hancock et al., 2022; Huang, 2017; Saiphoo et al., 2020).

So, in the next chapter, I want to examine the questions: *When do we diminish machines?* Or more specifically, *What factors cause the machine penalty, or, at least, make it more likely?*

To answer these questions, I'll talk about specific studies, citing the conclusions of the systematic literature reviews and the meta-analysis, which are drawing conclusions based on their analysis of the state-of-the-art research in this area. From these, I highlight ten different causes of the machine penalty: five human–AI comparisons and five situations.

References

Andrews, E. (2023 [2015]). Who Were the Luddites? *History.com*. https://www.history.com/news/who-were-the-luddites

Baumeister, R. F., Bratslavsky, E., Finkenauer, C., & Vohs, K. D. (2001). Bad is stronger than good. *Review of General Psychology, 5*(4), 323–370. https://doi.org/10.1037/1089-2680.5.4.323

Brahambhatt, R. (2022). Mechanical Turk: An elaborate 18th-century hoax that played chess like an AI robot. *BigThink*. https://bigthink.com/the-past/mechanical-turk/

Burton, J. W., Stein, M. K., & Jensen, T. B. (2020). A systematic review of algorithm aversion in augmented decision making. *Journal of Behavioral Decision Making, 33*(2), 220–239. https://doi.org/10.1002/bdm.2155

Chandrasekaren, R. (1997). Kasparov Proves No Match for Computer. *The Washington Post.* https://www.washingtonpost.com/wp-srv/tech/analysis/kasparov/kasparov.htm

Conniff, R. (2011). What the luddites really fought against. *Smithsonian Magazine.* https://www.smithsonianmag.com/history/what-the-luddites-really-fought-against-264412/

Dawes, R. M. (1979). The robust beauty of improper linear models in decision making. *American Psychologist, 34*(7), 571–582. https://doi.org/10.1037/0003-066X.34.7.571

Dietvorst, B. J., Simmons, J. P., & Massey, C. (2015). Algorithm aversion: People erroneously avoid algorithms after seeing them err. *Journal of Experimental Psychology: General, 144*(1), 114–126. https://doi.org/10.1037/xge0000033

Garvey, A. M., Kim, T., & Duhachek, A. (2023). Bad news? Send an AI. Good news? Send a human. *Journal of Marketing, 87*(1), 10–25. https://doi.org/10.1177/00222429211066972

Hancock, J. T., Liu, S. X., Luo, M., & Mieczkowski, H. (2022). Social media and psychological well-being. In *The psychology of technology: Social science research in the age of Big Data* (pp. 195–238). American Psychological Association. https://doi.org/10.1037/0000290-007

Huang, C. (2017). Time spent on social network sites and psychological well-being: A meta-analysis. *Cyberpsychology, Behavior, and Social Networking, 20*(6), 346–354. https://doi.org/10.1089/cyber.2016.0758

Jussupow, E., Benbasat, I., & Heinzl, A. (2020). *Why are we averse towards algorithms? A comprehensive literature review on algorithm aversion.* Twenty-Eighth European Conference on Information Systems (ECIS), An Online AIS Conference, Virtual. https://aisel.aisnet.org/ecis2020_rp/168

Logg, J. M., Minson, J. A., & Moore, D. A. (2019). Algorithm appreciation: People prefer algorithmic to human judgment. *Organizational Behavior and Human Decision Processes, 151,* 90–103. https://doi.org/10.1016/j.obhdp.2018.12.005

Mahmud, H., Islam, A. N., Ahmed, S. I., & Smolander, K. (2022). What influences algorithmic decision-making? A systematic literature review on algorithm aversion. *Technological Forecasting and Social Change, 175,* 121390. https://doi.org/10.1016/j.techfore.2021.121390

Meehl, P. E. (1954). *Clinical versus statistical prediction: A theoretical analysis and a review of the evidence*. University of Minnesota Press. https://doi.org/10.1037/11281-000

Morewedge, C. K. (2009). Negativity bias in attribution of external agency. *Journal of Experimental Psychology: General, 138*(4), 535–545. https://doi.org/10.1037/a0016796

Qin, X., Zhou, X., Chen, C., Wu, D., Zhou, H., Dong, X., et al. (in press). AI Aversion or Appreciation? A Capability-Personalization Framework and a Meta-Analytic Review. *Psychological Bulletin*.

Saiphoo, A. N., Halevi, L. D., & Vahedi, Z. (2020). Social networking site use and self-esteem: A meta-analytic review. *Personality and Individual Differences, 153*, 109639. https://doi.org/10.1016/j.paid.2019.109639

Shank, D. B. (2008). *Affect toward computers who coerce in social exchange*. Master's thesis, University of Georgia, Athens, GA. https://getd.libs.uga.edu/pdfs/shank_daniel_b_200805_ma.pdf

Shank, D. B. (2012a). *Agency and emotion in interactions with technological representatives of organizations*. Doctoral dissertation, University of Georgia, Athens, GA. https://esploro.libs.uga.edu/esploro/outputs/doctoral/Agency-and-emotion-in-interactions-with/9949334258002959

Shank, D. B. (2012b). Perceived justice and reactions to coercive computers. *Sociological Forum, 27*(2), 372–391. https://doi.org/10.1111/j.1573-7861.2012.01322.x

Shank, D. B. (2013). Are computers good or bad for business? How mediated customer–computer interaction alters emotions, impressions, and patronage toward organizations. *Computers in Human Behavior, 29*(3), 715–725. https://doi.org/10.1016/j.chb.2012.11.006

Shank, D. B. (2014). Impressions of computer and human agents after interaction: Computer identity weakens power but not goodness impressions. *International Journal of Human-Computer Studies, 72*(10-11), 747–756. https://doi.org/10.1016/j.ijhcs.2014.05.002

Stephens, E. (2023). The mechanical Turk: A short history of 'artificial artificial intelligence'. *Cultural Studies, 37*(1), 65–87. https://doi.org/10.1080/09502386.2022.2042580

6

The Machine Penalty Causes

There are many potential ways we might diminish machines compared to humans, based on the diversity of machines, humans, and interaction situations. AI comes in all forms from embodied robots to faceless programs, with different levels of sophistication and even different levels of name recognition. They engage in all sorts of decision making, advising, creating, evaluating, and interacting with humans, and produce outcomes of all types.

Humans, too, are not all the same, from children to adults and novices to experts. And comparisons made between humans and AI vary considerably based on which humans and which AI (*AI vs Human Condition*) as well as the situation under which the comparison is being made (*Situational Effect Condition*). This last sentence not only captures the complexity and range of cases in which an outcome could be diminished when they come from an AI versus a human but also offers a way to classify the primary causes of the machine penalty: human–AI comparisons and situations (Fig. 1.2).

In this chapter, I cover how the five human–AI comparisons and five situations can cause the machine penalty. We must first differentiate between humans and AI, or at least between an individual human and an

D. B. Shank, *The Machine Penalty*, https://doi.org/10.1007/978-3-031-86061-4_6

individual AI, or there is no chance we can penalize the AI (*AI vs Human Condition*). That's why human–AI comparisons are the root causes. Yet, the specific situation is often just as—or even more—critical in determining whether we penalize the AI or not (*Situational Effect Condition*).

6.1 Human–AI Comparisons

In Chap. 4, I laid a foundation by considering how appearance, identity, behavior, mind, and essence comparisons differentiated humans and AI. AI, individually or as a collective, fails to reach the human benchmark in these areas, sometimes with one comparison triggering other comparisons. Now, let's consider what the research shows regarding how each of these comparisons leads to the machine penalty.

6.1.1 Appearance

We tend to anthropomorphize and perceive computers as social actors, leading to us treating machines with humanlike physical features more like humans. Therefore, increased anthropomorphism and anthropomorphic design typically reduces the machine penalty (Ahn et al., 2021), meaning less anthropomorphic machines are more frequently penalized.

We're more trusting and accepting of AI with a humanlike appearance including anthropomorphized autonomous vehicles (Waytz et al., 2014), AI using a human headshot photo (Pak et al., 2012) or avatar (Qiu & Benbasat, 2009), or robots (Zhao & Malle, 2022). Additionally, voice leads us to perceive machines more like humans (Schroeder & Epley, 2016) with a more personalized rather than mechanical conversation style increasing acceptance of AI (Yun et al., 2021).

One reason that anthropomorphized appearance decreases the machine penalty is that with anthropomorphism comes more social presence, which increases one's closeness to and acceptance of the AI (Mahmud et al., 2022; Qiu & Benbasat, 2009). However, there are cases when a more human-like appearance does not give us more positive interactions with an AI. Namely, when a machine is in the uncanny valley, the benefits of more humanlike appearance is offset by the uncomfortable eeriness people have interacting with them (Laakasuo et al., 2021).

Based on appearance, the machine penalty increases when AI looks less like humans or when machines appear uncannily close to humans.

6.1.2 Identity

The machine penalty is often present in situations wherein the only observable human–machine difference is their labeled identities, not appearances, nor behavior. But the way we label AI, and the humans they are compared to (*AI vs Human Condition*), are paramount to how we evaluate and potentially diminish them (Mariadassou et al., 2024). For example, a smart speaker identifying itself as a friend compared to an assistant makes people feel more positively toward it (Kim et al., 2019).

We perceive machine labels such as *software program*, *bot*, *artificial intelligence*, *digital interface*, *robot*, or *device* as indicative of different levels of power, goodness, and even mind (Shank et al., 2020). These differences persist when compared to humans. For example, when I compared across a range of different types of roles—like crane operators, tailors, Ponzi scammers, and hackers—machines labeled as *artificial intelligence* and *computer systems* were perceived as more powerful than humans in those roles, demonstrating a machine advantage for power. However, *artificial intelligence* was also perceived as less positive than both humans and *computer systems,* demonstrating a machine penalty for positive impressions of machines labeled as *artificial intelligence* (Shank, Bowen, et al., 2021).

Perhaps the most important identity for humans or machines is whether they are an expert regarding the task at hand, which is an identity that indicates expected behavior. People or AI being identified as experts increases the likelihood of following either (Jussupow et al., 2020), but expertise gives an advantage to humans. While novice AI are trusted over novice humans, expert humans are generally trusted over expert AI (Madhavan & Wiegmann, 2007). The exception is when AI's expertise in a domain will clearly be superior (Pearson et al., 2019), either based on past performance or knowledge of the AI's abilities. Not surprisingly, expert AI are trusted and followed more than novice humans (Gunaratne et al., 2018).

Based on identity, the machine penalty increases for specific machine labels or when AI are compared to human experts.

6.1.3 Behavior

Like the findings for the identity of "expert," if a human or AI demonstrates expertise and outperforms the other, we generally prefer the superior performer. This superior behavior gives us an indication of abilities, and, therefore, future behavior and outcomes.

However, even similar past behavior is judged differently with machine errors, failures, and poor performance as stronger indicators of future performance and abilities (Jussupow et al., 2020; Mahmud et al., 2022), sometimes due to our expectation that the machine should have flawless behavior (Dzindolet et al., 2002). This explains results, like the original algorithm aversion study (Dietvorst et al., 2015) and subsequent research (Renier et al., 2021), that finds a machine penalty for AI even with superior performance if the AI demonstrates failures or errors. Since people often assume that AI are unable to learn from their mistakes (Dietvorst et al., 2015; Reich et al., 2023), behavior that displays the AI's learning ability reduces the machine penalty (Berger et al., 2021).

Expectedly, repeated past AI behavior leads to familiarity, which—as long as it's accompanied by a lack of bad experiences—sets up positive expectations, thereby decreasing the machine penalty (Berger et al., 2021; Mahmud et al., 2024; Mahmud et al., 2022). Clearly past quality behavior increases our expectations, but this research shows that it does so more for AI compared to humans. Yet being more familiar with an AI's previously error-prone behavior leads to a greater machine penalty (Prahl & Van Swol, 2017). Interestingly, just by an AI simulating the events leading to an outcome—a process more familiar to us than machine learning and statistical methods—increases our preference for it (Fei & Dietvorst, 2023).

Similar AI and human behavior can also be judged differently. The same cheating behaviors make AI seem more powerful than humans, whereas the same non-cheating behaviors make humans seem more moral than AI (Shank et al., 2024). A classic AI study used a rudimentary psychotherapy chatbot named Eliza to parrot back phases in a text-only counseling session. When people thought Eliza was human, they reported the therapy was more helpful than when they were told it was a program (Weizenbaum, 1966).

The speed of behavior has an interesting but opposite effect, whereby fast predictions by AI, but slow ones by humans, are perceived as higher-quality and lead people to increasingly value and follow them (Efendić et al., 2020). However, a slower response from an AI also allows for the perceiver to spend more time analyzing the task at hand and, therefore, prepare them to better utilize the AI response (Park et al., 2019).

Based on behavior, the machine penalty increases primarily when AI make mistakes, have errors, or fail to demonstrate they can learn.

6.1.4 Mind

We tend to perceive robots and AI as having less agency and experience than humans (Gray et al., 2007; Wegner & Gray, 2017). Yet our perception of AI minds—both agency and experience—can change based on the AI's appearance (Appel et al., 2020; Gray & Wegner, 2012; Martini et al., 2016; Schroeder & Epley, 2016; Zhao et al., 2019), identity (Eyssel & Kuchenbrandt, 2012; Waytz et al., 2014), behavior (Bigman & Gray, 2018; Gray & Wegner, 2012; Shank & DeSanti, 2018; Waytz, Morewedge, et al., 2010), and situation (Tanibe et al., 2017; Ward et al., 2013). Nonetheless, these variations rarely cause us to perceive a machine as a fully minded human, which suggests that we perceive some essential differences in the mind of humans, compared to machines.

We often diminish AI compared to humans based on our perception they have less mind, and differences in mind often account for the diminished outcomes. For example, mind perception differences explain why we prefer humans to make moral decisions (Bigman et al., 2023), how much we perceive AI have virtuous or unvirtuous character (Shank,

North, et al., 2021), how much we want to use robots for eldercare (Stafford et al., 2014), and how much we buy from machines (Luo et al., 2019).

Likewise, in their book of studies on how human and AI outcomes are judged differently, Cesar Hidalgo and colleagues concluded with two principles related to intentions, which is a major part of having agency, or an agentic mind. First, we judge humans more by intentions and machines more by outcomes. Second, we assign more varied intentions to humans and more narrow intentions to machines (Hidalgo et al., 2021). In sum, perceived intentional mind causes the machine penalty.

In many ways mind perception is the link between appearance and other physical causes to more essentialist beliefs and expectations about AI including what they can feel and think. If they can feel and think like humans—assuming we aren't viewing them as eerily uncanny (Gray & Wegner, 2012; Stein et al., 2020)—then we tend to judge them more equally in terms of moral rights and responsibilities (Gray et al., 2012; Waytz, Gray, et al., 2010) and social closeness (Lee et al., 2020).

Intriguingly, we believe that we understand the mind and decision-making processes of other humans better than AI, because we can project our own thinking and feeling processes onto other people. While this might be true, we tend to overestimate our understanding of other people's internal processes. Therefore, one reason we discount machine behavior compared to humans is that we perceive greater understanding of other people than is actually warranted (Bonezzi et al., 2022).

Based on mind, the machine penalty increases when AI are perceived to have lower agency or experience.

6.1.5 Essence

Most people believe that AI along with other machines have a different essential nature than humans, rooted in being biological or mechanical (Haslam et al., 2000). This leads to a number of assumed expectations for behavior (Burton et al., 2020), such as AI should have perfect performance (Dzindolet et al., 2002), can't learn (Dietvorst et al., 2015; Reich et al., 2023), maximize outcomes (Dietvorst & Bartels, 2022), fail to consider unique context or circumstances (Newman et al., 2020), have errors that are systematic (Dietvorst et al., 2016), and aren't empathetic to humans (Luo et al., 2019).

If we hold on to those expectations, then any behaviors that point to errors or deficiencies make us quicker to diminish the AI, compared to similar errors by humans (Dzindolet et al., 2002). Yet people also expect AI to have more consistent behaviors, and humans to have more variable ones, suggesting that collectively humans might both have more inferior and more superior behavior compared to AI (Dietvorst & Bharti, 2020).

Other expectations about the essential nature of AI mean that we might assume incompatibilities with how an AI arrives at an outcome, using probability and logic instead of intuition (Burton et al., 2020). Specifically, AI often operates as a "black box" without transparency or explanations as to its decisions. Not only does that leave perceivers to make their own assumptions, but people are more likely to exhibit a machine penalty when they do not understand how the AI reached its decision (Mahmud et al., 2022), even if they are pleased with the decision (Yeomans et al., 2019). In short, not being able to understand AI's reasoning, and presuming we understand humans' reasoning (Bonezzi et al., 2022), leads to a decrease in trust (Goodwin et al., 2013; van Dongen & van Maanen, 2013), which often manifests as the machine penalty.

Based on essence, the machine penalty increases when AI are believed to lack specific human abilities or capacities.

6.2 Situations

Situations not only help reveal and enable human–AI comparisons, which cause the machine penalty, but often involve aspects paramount to determining whether the penalty will occur and to what degree. These five situational causes range across focusing on the perceiver, the decision, and the environmental context.

6.2.1 Controllable

We like to be involved in decisions and have a hard time giving those up. Even those of us who are not control freaks don't want to let someone or something take over relevant, important, and novel decisions. In general, we don't want to lose our autonomy or have situations outside of our control (Glass et al., 1971; Langer, 1975).

Part of this is that we are biased toward ourselves: we tend to inflate our estimates of our own abilities (Shepperd et al., 2008; Yaniv & Kleinberger, 2000), discount recommendations of others compared to ourselves (Yaniv & Kleinberger, 2000), and even believe we will be luckier than others (Weinstein, 1989). Of course, there are some decisions and activities that we are happy to outsource, especially if they are trivial, unpleasant, or time-consuming, but usually we desire to keep a modicum of control even over those.

The machine penalty is greater for those who initially have the most control. Perceivers who are experts in an area, have strong preferences, or have greater confidence in their own decision-making penalize AI more than humans (Burton et al., 2020; Mahmud et al., 2022). Our desire for controllable situations means we will not easily cede decisions, especially important or relevant decisions, to AI or other humans. In situations where we generally would prefer an AI over another human—cases of the machine advantage—this advantage is suppressed and even reversed if the other human is yourself (Logg et al., 2019).

As opposed to turning over control, when we receive advice, we remain in control because we can always listen and then decide not to follow it, following our own preferences (Elder et al., 2024). Because of this, a small machine penalty for advice grows into a large machine penalty for fully turning over a decision to an AI or human (Dietvorst et al., 2015; Jussupow et al., 2020). Even when receiving AI advice, we prefer to control the details and complexity of that AI advice (Subramanian et al., 2024).

It is also possible to preserve our autonomy in some situations, such as if we are allowed to modify or test the AI, in which case the penalty is often reduced or eliminated (Burton et al., 2020; Dietvorst et al., 2016; Mahmud et al., 2022; van Dongen & van Maanen, 2013). Even when the accuracy of the AI is kept constant, being allowed to select the type of error that an AI should avoid or commit increases our acceptance of it (Kocielnik et al., 2019).

Based on controllability, the machine penalty increases as the perceiver loses autonomy or desires more control.

6.2.2 Personal

When situations and their outcomes are personally relevant or personalized to us, we have increased investment in that situation and believe the outcome needs to be uniquely specific. In fact, this ties to the previous cause in that if it is personally important, then people desire more control. If it is personally important, then you'll also carefully think whether an AI or human is best in that situation, which can lead to a different choice based on other factors. In general, the machine penalty weakens and we rely on AI more when we are more personally responsible for a decision (van Dongen & van Maanen, 2013).

Analyzing 163 studies on preference for AI or humans, Qin et al.'s meta-analysis shows that situations that require personalization overwhelmingly support the machine penalty (Qin et al., in press). Namely, we prefer humans in personal situations. This is driven, at least in part, by the expectation that AI can't account for unique characteristics of individuals (Longoni et al., 2019; Newman et al., 2020) and therefore are bad at personalization (Longoni & Cian, 2020). When AI are framed as being capable of personalizing health advice, people accept that advice at similar levels to human advice (Longoni et al., 2019).

Likewise if the situation threatens a personal identity, we are more likely to penalize machines compared to humans, especially in situations with more ambiguous criteria and outcomes (Morewedge, 2022). For example, some people really enjoy driving a car or have an identity as a coffee barista. These levels of enjoyment and identification will result in stronger machine penalties regardless of one's beliefs about the safety of autonomous vehicles or the quality of lattes produced by an espresso machine.

Based on the personal, the machine penalty increases when the outcome is of personal importance or is personalized for the perceiver.

6.2.3 Important

In general, the machine penalty weakens, and we rely on AI more when there is an incentive to reach a correct decision and when the decision is important in general (Köbis & Mossink, 2021; Mahmud et al., 2022). Likewise, we tend to prefer AI more when making actual judgments and discount them more in hypothetical judgments where we often prefer humans over AI (Logg & Schlund, 2024). Of course, these are only in cases where we expect the AI to produce a high-quality outcome that is similar to a human.

In cases where we believe AI could be capable of making a good decision, we depend on it more when we are personally responsible for that decision (van Dongen & van Maanen, 2013). That is, the importance of personal accountability weakens the machine penalty. Thus, overall, we see the penalty operating less for more consequential and important decisions we are accountable for, and thus operating more for trivial and mundane decisions with no such accountability.

Essentially, these are cases where we go through a slower, more critical thinking process where we realize the AI could produce a quality—or even superior—outcome and therefore decide we should prefer it. If we come to the conclusion that the AI is ill-suited, then the situation's importance contributes to the penalty. Importance, therefore, can enhance existing preferences whether that be toward the AI or the human. Of course, if we see actual output from AI that is superior to human output, then we prefer its outcomes based on that, but if the output is similar or we've yet to see, importance generally leads to a machine advantage.

The machine advantage for important decisions is at odds with the machine penalty for personal decisions. The resolution of this depends on what way a decision is important. If a decision is weightier in general but is not personally important for the perceiver, we'll tend to give added advantage to the machine. But, if its importance comes only from the perceiver's priorities and connection to their life and interests, we'll penalize the machine as discussed in the previous section.

Based on importance, the machine penalty increases when the outcome is mundane or less important in general.

6.2.4 Subjective

The penalty is stronger and found more often for hedonistic, empathic, personalized, subjective decisions, whereas the machine advantage is stronger and more often found for quantitative, objective, utilitarian, and logical decisions (Castelo et al., 2019; Jussupow et al., 2020; Longoni & Cian, 2020; Mahmud et al., 2022).

This is tied to the essence of machines, including their computational nature, lack of a human mind, and lack of emotions and human experiences. This also relates to expected behaviors from a machine, such as being good at mathematical and quantitative tasks. We are therefore more trusting of machines for objective quantification, and less trusting for subjective evaluations, at least in comparison to humans. Humans, of course, do both, but arguably have to train and exert effort to be more objective and quantitative.

Interestingly, tasks might seem like they would clearly be subjective or objective, but there is a continuum and within a situation our perception of the subjectivity or objectivity of a task can be shifted (Castelo et al., 2019). By doing so, Castelo and colleagues showed as people believed that a task is more objective, that they are more likely to trust and use AI for that task (Castelo et al., 2019), as perceived objectivity leads to the machine advantage (Zhang et al., 2021).

This difference in objectivity or subjectivity might also be present in the same situation, but for different aspects of the outcomes. For example, an AI recommendation of a coat based on its fashionability—a more subjective evaluation—is diminished more compared to its recommendation of the same coat based on its warmth—a more objective criterion. Therefore, there is a machine penalty for the coat recommendation based on fashionability, but a machine advantage for the coat recommendation based on its utility (Longoni & Cian, 2020).

Based on subjectivity, the machine penalty increases when the situation involves subjective, hedonistic, or empathic decisions.

6.2.5 Moral

We discount AI outcomes related to moral choices and moral decisions that affect humans' wellbeing. This means the machine penalty is often observed for ethical, legal, military, safety, or medical decisions (Bigman & Gray, 2018; Mahmud et al., 2022), especially when those decisions have important consequences (Castelo et al., 2019). In some cases AI have been shown to make superior ethical decisions compared to humans (Dillion et al., 2024), whereas in other cases, they've been shown to give suspect and inconsistent moral advice (Krügel et al., 2023).

When given the choice, we tend to select humans, rather than AI, to make a range of moral decisions in a range of different contexts (Bigman & Gray, 2018). In these cases, the penalty is often due to our beliefs about the mind of AI, that they don't feel or think in the same way as humans, and, therefore, will be lacking in critical ethical judgments with real-world consequences (Bigman & Gray, 2018; Shank, North, et al., 2021). Currently, we tend to think that AI are outside the circle of agents, which should be given moral consideration (Pauketat & Anthis, 2022), making them seem less in tune to moral concerns. Moral decisions may also be more subjectively based on one's personal and ethical beliefs, and, therefore, the same arguments on subjective and personal situations may also apply to moral ones.

After immoral actions have already occurred, we also diminish outcomes for AI compared to humans. People also judge identical unethical outcomes produced by an AI as less wrong, less blameworthy, and are less outraged by them than when they are produced by a human (Bigman et al., 2023; Maninger & Shank, 2022). These judgments hold across different types of moral violations (Maninger & Shank, 2022), including real-world moral violations originally perpetrated by AI, not humans (Shank, DeSanti, et al., 2019; Wilson et al., 2022).

Based on morality, the machine penalty increases when the situation involves decisions dealing with morality or human welfare.

6.3 Summary of Causes

These ten causes capture the primary reasons why the machine penalty occurs and under what conditions it occurs. To summarize, Table 6.1 overviews all the causes, with a brief explanation of each of the causes most substantial effect on the machine penalty.

Let me end with three notes and caveats on my conceptualization of the causes.

First, others have different arguments on what constitutes the most vital causes of the machine penalty (Castelo et al., 2019; Jussupow et al., 2020; Morewedge, 2022), although their conceptualizations overlap or are included in the causes just presented.

Second, I selected these ten as the most relevant causes, to be both comprehensive but also to keep it manageable. As such, sometimes I have combined causes that others might separate. For example, I included roles and labels both within identity comparisons.

Third, there are also other causes that I find less applicable because they are individual differences between perceivers such as personality characteristics or general aversion to AI and machines (Grundke et al., 2024) or demographic differences (Hidalgo et al., 2021). Similarly, some

Table 6.1 Ten causes of the machine penalty

Cause	Cause type	Machine penalty increases when
Appearance	Comparison	AI look less like humans or appear uncanny
Identity	Comparison	AI are compared to human experts
Behavior	Comparison	AI make mistakes or are imperfect
Mind	Comparison	AI have lower perceived agency or experience
Essence	Comparison	AI are believed to lack some human ability or capacity
Controllable	Situation	the perceiver desires more control or may lose control
Personal	Situation	the outcome is personal or personalized for the perceiver
Important	Situation	the outcome is mundane or less important in general
Subjective	Situation	the situation involves subjective decisions
Moral	Situation	the situation involves moral or human welfare decisions

causes might be contextual or framing effects. For example, thinking about God leads to an AI advantage because that makes us feel small as humans and view humans as more fallible, therefore lowering our dependence on humans (Karataş & Cutright, 2023). That situational effect, while interesting, doesn't neatly fit in these cause categories. If you're interested in a comprehensive list of other causes, I recommend the systematic literature review and classification of causes by Hasan Mahmud et al. (2022).

In conclusion, in this chapter and throughout Part I, I've presented the machine penalty as not only a simple idea to understand but also a more complex idea in understanding when it will occur. The ten causes help pin down when it is the most likely to occur and when it most likely will not occur. In Part II, we'll consider its application to advice, evaluations, harm, art, and companionship. For those, we'll see how it connects to these ten causes, and, because of them, operates a bit differently in each domain.

References

Ahn, J., Kim, J., & Sung, Y. (2021). AI-powered recommendations: The roles of perceived similarity and psychological distance on persuasion. *International Journal of Advertising, 40*(8), 1366–1384. https://doi.org/10.1080/02650487.2021.1982529

Appel, M., Izydorczyk, D., Weber, S., Mara, M., & Lischetzke, T. (2020). The uncanny of mind in a machine: Humanoid robots as tools, agents, and experiencers. *Computers in Human Behavior, 102*, 274–286. https://doi.org/10.1016/j.chb.2019.07.031

Berger, B., Adam, M., Rühr, A., & Benlian, A. (2021). Watch me improve – Algorithm aversion and demonstrating the ability to learn. *Business & Information Systems Engineering, 63*(1), 55–68. https://doi.org/10.1007/s12599-020-00678-5

Bigman, Y. E., & Gray, K. (2018). People are averse to machines making moral decisions. *Cognition, 181*, 21–34. https://doi.org/10.1016/j.cognition.2018.08.003

Bigman, Y. E., Wilson, D., Arnestad, M. N., Waytz, A., & Gray, K. (2023). Algorithmic discrimination causes less moral outrage than human discrimination. *Journal of Experimental Psychology: General, 152*(1), 4–27. https://doi.org/10.1037/xge0001250

Bonezzi, A., Ostinelli, M., & Melzner, J. (2022). The human black-box: The illusion of understanding human better than algorithmic decision-making. *Journal of Experimental Psychology: General, 151*(9), 2250–2258. https://doi.org/10.1037/xge0001181

Burton, J. W., Stein, M. K., & Jensen, T. B. (2020). A systematic review of algorithm aversion in augmented decision making. *Journal of Behavioral Decision Making, 33*(2), 220–239. https://doi.org/10.1002/bdm.2155

Castelo, N., Bos, M. W., & Lehmann, D. R. (2019). Task-dependent algorithm aversion. *Journal of Marketing Research, 56*(5), 809–825. https://doi.org/10.1177/0022243719851788

Dietvorst, B. J., & Bartels, D. M. (2022). Consumers object to algorithms making morally relevant tradeoffs because of algorithms' consequentialist decision strategies. *Journal of Consumer Psychology, 32*(3), 406–424. https://doi.org/10.1002/jcpy.1266

Dietvorst, B. J., & Bharti, S. (2020). People reject algorithms in uncertain decision domains because they have diminishing sensitivity to forecasting error. *Psychological Science, 31*(10), 1302–1314. https://doi.org/10.1177/0956797620948841

Dietvorst, B. J., Simmons, J. P., & Massey, C. (2015). Algorithm aversion: People erroneously avoid algorithms after seeing them err. *Journal of Experimental Psychology: General, 144*(1), 114–126. https://doi.org/10.1037/xge0000033

Dietvorst, B. J., Simmons, J. P., & Massey, C. (2016). Overcoming algorithm aversion: People will use imperfect algorithms if they can (even slightly) modify them. *Management Science, 64*(3), 1155–1170. https://doi.org/10.1287/mnsc.2016.2643

Dillion, D., Mondal, D., Tandon, N., & Gray, K. (2024). *Large language models as moral experts? GPT-4o outperforms expert ethicist in providing moral guidance*. https://doi.org/10.31234/osf.io/w7236

Dzindolet, M. T., Pierce, L. G., Beck, H. P., & Dawe, L. A. (2002). The perceived utility of human and automated aids in a visual detection task. *Human Factors, 44*(1), 79–94. https://doi.org/10.1518/0018720024494856

Efendić, E., Van de Calseyde, P. P. F. M., & Evans, A. M. (2020). Slow response times undermine trust in algorithmic (but not human) predictions. *Organizational Behavior and Human Decision Processes, 157*, 103–114. https://doi.org/10.1016/j.obhdp.2020.01.008

Elder, H., Canfield, C., Shank, D. B., Rieger, T., & Hines, C. (2024). Knowing when to pass: The effect of AI reliability in risky decision contexts. *Human Factors, 66*(2), 348–362. https://doi.org/10.1177/00187208221100691

Eyssel, F., & Kuchenbrandt, D. (2012). Social categorization of social robots: Anthropomorphism as a function of robot group membership. *British Journal of Social Psychology, 51*(4), 724–731. https://doi.org/10.1111/j.2044-8309.2011.02082.x

Fei, L., & Dietvorst, B. (2023). Prediction by replication: People prefer prediction algorithms that replicate the event being predicted. *Journal of the Association for Consumer Research, 9*(3), 306–318. https://doi.org/10.1086/728289

Glass, D. C., Reim, B., & Singer, J. E. (1971). Behavioral consequences of adaptation to controllable and uncontrollable noise. *Journal of Experimental Social Psychology, 7*(2), 244–257. https://doi.org/10.1016/0022-1031(71)90070-9

Goodwin, P., Sinan Gönül, M., & Önkal, D. (2013). Antecedents and effects of trust in forecasting advice. *International Journal of Forecasting, 29*(2), 354–366. https://doi.org/10.1016/j.ijforecast.2012.08.001

Gray, H. M., Gray, K., & Wegner, D. M. (2007). Dimensions of mind perception. *Science, 315*(5812), 619. https://doi.org/10.1126/science.1134475

Gray, K., & Wegner, D. M. (2012). Feeling robots and human zombies: Mind perception and the uncanny valley. *Cognition, 125*(1), 125–130. https://doi.org/10.1016/j.cognition.2012.06.007

Gray, K., Young, L., & Waytz, A. (2012). Mind perception is the essence of morality. *Psychological Inquiry, 23*(2), 101–124. https://doi.org/10.1080/1047840X.2012.651387

Grundke, A., Appel, M., & Stein, J.-P. (2024). Aversion against machines with complex mental abilities: The role of individual differences. *Computers in Human Behavior: Artificial Humans, 2*(2), 100087. https://doi.org/10.1016/j.chbah.2024.100087

Gunaratne, J., Zalmanson, L., & Nov, O. (2018). The persuasive power of algorithmic and crowdsourced advice. *Journal of Management Information Systems, 35*(4), 1092–1120. https://doi.org/10.1080/07421222.2018.1523534

Haslam, N., Rothschild, L., & Ernst, D. (2000). Essentialist beliefs about social categories. *British Journal of Social Psychology, 39*(1), 113–127. https://doi.org/10.1348/014466600164363

Hidalgo, C. A., Orghian, D., Canals, J. A., De Almeida, F., & Martin, N. (2021). *How humans judge machines*. MIT Press.

Jussupow, E., Benbasat, I., & Heinzl, A. (2020). *Why are we averse towards algorithms? A comprehensive literature review on algorithm aversion*. Twenty-Eighth European Conference on Information Systems (ECIS), An Online AIS Conference, Virtual. https://aisel.aisnet.org/ecis2020_rp/168

Karataş, M., & Cutright, K. M. (2023). Thinking about God increases acceptance of artificial intelligence in decision-making. *Proceedings of the National Academy of Sciences, 120*(33), e2218961120. https://doi.org/10.1073/pnas.2218961120

Kim, A., Cho, M., Ahn, J., & Sung, Y. (2019). Effects of gender and relationship type on the response to artificial intelligence. *Cyberpsychology, Behavior, and Social Networking, 22*(4), 249–253. https://doi.org/10.1089/cyber.2018.0581

Köbis, N., & Mossink, L. D. (2021). Artificial intelligence versus Maya Angelou: Experimental evidence that people cannot differentiate AI-generated from human-written poetry. *Computers in Human Behavior, 114*, 106553. https://doi.org/10.1016/j.chb.2020.106553

Kocielnik, R., Amershi, S., & Bennett, P. N. (2019). *Will you accept an imperfect AI? Exploring designs for adjusting end-user expectations of AI systems* Proceedings of the 2019 CHI Conference on Human Factors in Computing Systems, Glasgow, Scotland UK. https://doi.org/10.1145/3290605.3300641

Krügel, S., Ostermaier, A., & Uhl, M. (2023). ChatGPT's inconsistent moral advice influences users' judgment. *Scientific Reports, 13*(1), 4569. https://doi.org/10.1038/s41598-023-31341-0

Laakasuo, M., Palomäki, J., & Köbis, N. (2021). Moral uncanny valley: A robot's appearance moderates how its decisions are judged. *International Journal of Social Robotics, 13*(7), 1679–1688. https://doi.org/10.1007/s12369-020-00738-6

Langer, E. J. (1975). The illusion of control. *Journal of Personality and Social Psychology, 32*(2), 311–328. https://doi.org/10.1037/0022-3514.32.2.311

Lee, S., Lee, N., & Sah, Y. J. (2020). Perceiving a mind in a chatbot: Effect of mind perception and social cues on co-presence, closeness, and intention to use. *International Journal of Human–Computer Interaction, 36*(10), 930–940. https://doi.org/10.1080/10447318.2019.1699748

Logg, J., & Schlund, R. (2024). A simple explanation reconciles "Algorithm Aversion" and "Algorithm Appreciation": Hypotheticals vs. real judgments. *Working Paper*. https://papers.ssrn.com/sol3/papers.cfm?abstract_id=4687557

Logg, J. M., Minson, J. A., & Moore, D. A. (2019). Algorithm appreciation: People prefer algorithmic to human judgment. *Organizational Behavior and Human Decision Processes, 151*, 90–103. https://doi.org/10.1016/j.obhdp.2018.12.005

Longoni, C., Bonezzi, A., & Morewedge, C. K. (2019). Resistance to medical artificial intelligence. *Journal of Consumer Research, 46*(4), 629–650. https://doi.org/10.1093/jcr/ucz013

Longoni, C., & Cian, L. (2020). Artificial intelligence in utilitarian vs. hedonic contexts: The "word-of-machine" effect. *Journal of Marketing, 86*(1), 91–108. https://doi.org/10.1177/0022242920957347

Luo, X., Tong, S., Fang, Z., & Qu, Z. (2019). Frontiers: Machines vs. humans: The impact of artificial intelligence Chatbot disclosure on customer purchases. *Marketing Science, 38*(6), 937–947. https://doi.org/10.1287/mksc.2019.1192

Madhavan, P., & Wiegmann, D. A. (2007). Effects of information source, pedigree, and reliability on operator interaction with decision support systems. *Human Factors, 49*(5), 773–785. https://doi.org/10.1518/001872007x230154

Mahmud, H., Islam, A. K. M. N., Luo, X., & Mikalef, P. (2024). Decoding algorithm appreciation: Unveiling the impact of familiarity with algorithms, tasks, and algorithm performance. *Decision Support Systems, 179*, 114168. https://doi.org/10.1016/j.dss.2024.114168

Mahmud, H., Islam, A. N., Ahmed, S. I., & Smolander, K. (2022). What influences algorithmic decision-making? A systematic literature review on algorithm aversion. *Technological Forecasting and Social Change, 175*, 121390. https://doi.org/10.1016/j.techfore.2021.121390

Maninger, T., & Shank, D. B. (2022). Perceptions of violations by artificial and human actors across moral foundations. *Computers in Human Behavior Reports, 5*, 100154. https://doi.org/10.1016/j.chbr.2021.100154

Mariadassou, S., Klesse, A.-K., & Boegershausen, J. (2024). Averse to what: Consumer aversion to algorithmic labels, but not their outputs? *Current Opinion in Psychology, 58*, 101839. https://doi.org/10.1016/j.copsyc.2024.101839

Martini, M. C., Gonzalez, C. A., & Wiese, E. (2016). Seeing minds in others – Can agents with robotic appearance have human-like preferences? *PloS ONE, 11*(1), e0146310. https://doi.org/10.1371/journal.pone.0146310

Morewedge, C. K. (2022). Preference for human, not algorithm aversion. *Trends in Cognitive Sciences, 26*(10), 824–826. https://doi.org/10.1016/j.tics.2022.07.007

Newman, D. T., Fast, N. J., & Harmon, D. J. (2020). When eliminating bias isn't fair: Algorithmic reductionism and procedural justice in human resource decisions. *Organizational Behavior and Human Decision Processes, 160*, 149–167. https://doi.org/10.1016/j.obhdp.2020.03.008

Pak, R., Fink, N., Price, M., Bass, B., & Sturre, L. (2012). Decision support aids with anthropomorphic characteristics influence trust and performance in younger and older adults. *Ergonomics, 55*(9), 1059–1072. https://doi.org/10.1080/00140139.2012.691554

Park, J. S., Barber, R., Kirlik, A., & Karahalios, K. (2019). A slow algorithm improves users' assessments of the algorithm's accuracy. In *Proceeding of the ACM on human-computer interaction* (Vol. 3, pp. Article 102). https://doi.org/10.1145/3359204

Pauketat, J. V. T., & Anthis, J. R. (2022). Predicting the moral consideration of artificial intelligences. *Computers in Human Behavior, 136*, 107372. https://doi.org/10.1016/j.chb.2022.107372

Pearson, C. J., Geden, M., & Mayhorn, C. B. (2019). Who's the real expert here? Pedigree's unique bias on trust between human and automated advisers. *Applied Ergonomics, 81*, 102907. https://doi.org/10.1016/j.apergo.2019.102907

Prahl, A., & Van Swol, L. (2017). Understanding algorithm aversion: When is advice from automation discounted? *Journal of Forecasting, 36*(6), 691–702. https://doi.org/10.1002/for.2464

Qin, X., Zhou, X., Chen, C., Wu, D., Zhou, H., Dong, X., et al. (in press). AI Aversion or Appreciation? A Capability-Personalization Framework and a Meta-Analytic Review. *Psychological Bulletin*.

Qiu, L., & Benbasat, I. (2009). Evaluating anthropomorphic product recommendation agents: A social relationship perspective to designing information systems. *Journal of Management Information Systems, 25*(4), 145–182. https://doi.org/10.2753/MIS0742-1222250405

Reich, T., Kaju, A., & Maglio, S. J. (2023). How to overcome algorithm aversion: Learning from mistakes. *Journal of Consumer Psychology, 33*(2), 285–302. https://doi.org/10.1002/jcpy.1313

Renier, L. A., Mast, M. S., & Bekbergenova, A. (2021). To err is human, not algorithmic–Robust reactions to erring algorithms. *Computers in Human Behavior, 124*, 106879. https://doi.org/10.1016/j.chb.2021.106879

Schroeder, J., & Epley, N. (2016). Mistaking minds and machines: How speech affects dehumanization and anthropomorphism. *Journal of Experimental Psychology: General, 145*(11), 1427–1437. https://doi.org/10.1037/xge0000214

Shank, D. B., Bowen, M., Burns, A., & Dew, M. (2021). Humans are perceived as better, but weaker, than artificial intelligence: A comparison of affective impressions of humans, AIs, and computer systems in roles on teams. *Computers in Human Behavior Reports, 3*, 100092. https://doi.org/10.1016/j.chbr.2021.100092

Shank, D. B., Burns, A., Rodriguez, S., & Bowen, M. (2020). Software program, bot, or artificial intelligence? Affective sentiments across general technology labels. *Current Research in Social Psychology, 28*(4), 32–41. https://crisp.org.uiowa.edu/sites/crisp.org.uiowa.edu/files/2020-06/crisp_28_4_shank.pdf

Shank, D. B., & DeSanti, A. (2018). Attributions of morality and mind to artificial intelligence after real-world moral violations. *Computers in Human Behavior, 86*, 401–411. https://doi.org/10.1016/j.chb.2018.05.014

Shank, D. B., DeSanti, A., & Maninger, T. (2019). When are artificial intelligence versus human agents faulted for wrongdoing? Moral attributions after individual and joint decisions. *Information, Communication, and Society, 22*(5), 648–663. https://doi.org/10.1080/1369118X.2019.1568515

Shank, D. B., Dew, M., & Sajjad, F. (2024). Moral behavior alters impressions of humans and AIs on teams: Unethical AIs are more powerful while ethical humans are nicer. *Behaviour & Information Technology*, 1–12. https://doi.org/10.1080/0144929X.2024.2403651

Shank, D. B., North, M., Arnold, C., & Gamez, P. (2021). Can mind perception explain virtuous character judgments of artificial intelligence? *Technology, Mind, and Behavior, 2*(3). https://doi.org/10.1037/tmb0000047

Shepperd, J., Malone, W., & Sweeny, K. (2008). Exploring causes of the self-serving bias. *Social and Personality Psychology Compass, 2*(2), 895–908. https://doi.org/10.1111/j.1751-9004.2008.00078.x

Stafford, R. Q., MacDonald, B. A., Jayawardena, C., Wegner, D. M., & Broadbent, E. (2014). Does the robot have a mind? Mind perception and attitudes towards robots predict use of an eldercare robot. *International Journal of Social Robotics, 6*(1), 17–32. https://doi.org/10.1007/s12369-013-0186-y

Stein, J.-P., Appel, M., Jost, A., & Ohler, P. (2020). Matter over mind? How the acceptance of digital entities depends on their appearance, mental prowess, and the interaction between both. *International Journal of Human-Computer Studies, 142*, 102463. https://doi.org/10.1016/j.ijhcs.2020.102463

Subramanian, H. V., Canfield, C., & Shank, D. B. (2024). Designing explainable AI to improve human-AI team performance: A medical stakeholder-driven scoping review. *Artificial Intelligence in Medicine, 149*, 102780. https://doi.org/10.1016/j.artmed.2024.102780

Tanibe, T., Hashimoto, T., & Karasawa, K. (2017). We perceive a mind in a robot when we help it. *PloS One, 12*(7), e0180952. https://doi.org/10.1371/journal.pone.0180952

van Dongen, K., & van Maanen, P.-P. (2013). A framework for explaining reliance on decision aids. *International Journal of Human-Computer Studies, 71*(4), 410–424. https://doi.org/10.1016/j.ijhcs.2012.10.018

Ward, A. F., Olsen, A. S., & Wegner, D. M. (2013). The harm-made mind: Observing victimization augments attribution of minds to vegetative patients, robots, and the dead. *Psychological Science, 24*(8), 1437–1445. https://doi.org/10.1177/0956797612472343

Waytz, A., Gray, K., Epley, N., & Wegner, D. M. (2010). Causes and consequences of mind perception. *Trends in Cognitive Sciences, 14*(8), 383–388. https://doi.org/10.1016/j.tics.2010.05.006

Waytz, A., Heafner, J., & Epley, N. (2014). The mind in the machine: Anthropomorphism increases trust in an autonomous vehicle. *Journal of Experimental Social Psychology, 52*, 113–117. https://doi.org/10.1016/j.jesp.2014.01.005

Waytz, A., Morewedge, C. K., Epley, N., Monteleone, G., Gao, J.-H., & Cacioppo, J. T. (2010). Making sense by making sentient: Effectance motivation increases anthropomorphism. *Journal of Personality and Social Psychology, 99*(3), 410–435. https://doi.org/10.1037/a0020240

Wegner, D. M., & Gray, K. (2017). *The mind club: Who thinks, what feels, and why it matters*. Penguin.

Weinstein, N. D. (1989). Optimistic biases about personal risks. *Science, 246*(4935), 1232–1233. https://doi.org/10.1126/science.2686031

Weizenbaum, J. (1966). ELIZA – A computer program for the study of natural language communication between man and machine. *Communications of the ACM, 9*(1), 36–45.

Wilson, A., Stefanik, C., & Shank, D. B. (2022). How do people judge the immorality of artificial intelligence versus humans committing moral wrongs in real-world situations? *Computers in Human Behavior Reports, 8*, 100229. https://doi.org/10.1016/j.chbr.2022.100229

Yaniv, I., & Kleinberger, E. (2000). Advice taking in decision making: Egocentric discounting and reputation formation. *Organizational Behavior and Human Decision Processes, 83*(2), 260–281. https://doi.org/10.1006/obhd.2000.2909

Yeomans, M., Shah, A., Mullainathan, S., & Kleinberg, J. (2019). Making sense of recommendations. *Journal of Behavioral Decision Making, 32*(4), 403–414. https://doi.org/10.1002/bdm.2118

Yun, J. H., Lee, E.-J., & Kim, D. H. (2021). Behavioral and neural evidence on consumer responses to human doctors and medical artificial intelligence. *Psychology & Marketing, 38*(4), 610–625. https://doi.org/10.1002/mar.21445

Zhang, L., Pentina, I., & Fan, Y. (2021). Who do you choose? Comparing perceptions of human vs robo-advisor in the context of financial services. *Journal of Services Marketing, 35*(5), 634–646. https://doi.org/10.1108/JSM-05-2020-0162

Zhao, X., & Malle, B. F. (2022). Spontaneous perspective taking toward robots: The unique impact of humanlike appearance. *Cognition, 224*, 105076. https://doi.org/10.1016/j.cognition.2022.105076

Zhao, X., Phillips, E., & Malle, B. F. (2019). *How people infer a humanlike mind from a robot body*. https://psyarxiv.com/w6r24/download

Part II

The Penalty Applied

7

Giving Advice

7.1 Are You Influenced?

The power of advice lies in its influence on people.

The news media reports dramatic stories when people followed machines' advice unquestioningly, showing us that AI's advice can influence people the wrong way. People drive off the road, sometimes to their deaths, because of faulty GPS directions (Hansen, 2015). And AI companions have encouraged people to commit suicide to stop climate change (El Atillah, 2023) and to assassinate Queen Elizabeth II (Associated Press, 2023).

While following AI's bad advice makes the biggest headlines, people routinely receive both good and bad advice from AI and from humans. So, the consequences depend not only on how much we're influenced to follow that advice but also whether it's advice worth following.

Stanislav Petrov might claim the world record for averting the worst possible consequences by ignoring machine advice. In 1983, at the height of the cold war, his military computers didn't recommend a specific decision, but provided information which, if believed, did suggest a clear course of action. The sensors and computers indicated to Petrov—a Lt.

D. B. Shank, *The Machine Penalty*, https://doi.org/10.1007/978-3-031-86061-4_7

Colonel in the Russian defense force—that the United States had launched five missiles at Russia. If he confirmed the threat and passed it up the chain-of-command, it would have easily sparked Russia's retaliation on the United States, probably beginning a nuclear war.

With maps and consoles flashing information and reports coming in over phones and intercoms, Petrov evaluated the machine's advice based on his familiarity with the machine, with likely military strategies of the United States, and with other data, and within five minutes concluded—correctly—that the report was false. In spite of the machine's clear indication of an attack, Petrov assessed that the machine was in error and potentially averted nuclear war (Hoffman, 1999).

Despite Petrov's choice having saved lives, diminishing or ignoring machine advice can also lead to poor and even catastrophic consequences. In 2009, Air France Flight 447 tragically crashed due to the combination of a thunderstorm, minor technical issues, cockpit crew communication, and mainly three pilots with little experience in understanding the aircraft's advanced computer systems.

The three pilots received an unambiguous warning from the system's automated voice, "Stall, Stall, Stall" indicating that the plane had, well, stalled. This implies a clear course of action for any trained pilot: point the plane's nose down and try to recover speed. Yet one of the pilots did the opposite, pointing the plane upward and putting on the brakes. This initially caused ascent, but subsequently the stalled plane slowed down which meant its altitude gain was short lived, and then reversed. As it began plunging toward the ocean the pilots remained confused, arguing about whether it was climbing or descending and whether it was stalled or not. By the time they realized the true situation, it was too late, and all 228 people on board died when they slammed into the ocean (Harford, 2016; Oliver et al., 2017).

That's the short version, but if we dig into some of the details, there was a much more complex interaction between the technology and the pilots. The automation system on the aircraft was designed to both fly the plane like an autopilot but also adjust for any maneuvers by the pilots that could be problematic—sort of like antilock brakes. This second feature, called fly-by-wire, was hailed as an advanced technological safeguard to keep pilots from accidentally crashing the plane, for example, by preventing extreme maneuvers that would stall it.

The entire incident began when ice from flying over a storm caused inconsistent speed sensor readings. This led to the automation system mainly disengaging itself, leaving the pilots in control with no autopilot but also with no fly-by-wire—no adjustments to their maneuvers. It was clear to them that there was no autopilot, but they didn't realize the fly-by-wire was mainly disengaged.

And they were not used to flying the aircraft without this automation, which set them up to ignore the warning advice from the computer. The pilots ignored the stall warning because they erroneously assumed the fly-by-wire part was active and would not allow them to stall (Harford, 2016). Therefore, the pilots refused to be influenced by this warning advice because of their expectations of the automation itself (Oliver et al., 2017; Schaich Borg et al., 2024). The strange irony of this tragedy was that too much trust in one computer system led them to ignore the obvious warning of another computer system.

The warning advice from the Flight 447 computer was correct, whereas the advice Petrov received from his military computer was wrong. Yet sometimes advice, especially in complex issues, can be both good and bad. For example, one American reporter planned her entire first-time visit to Norway using guidance from several different AI, including trip planning AI plus ChatGPT. She found that none of the AI advisors offered the complete guidance needed for the trip—often missing obvious good options in favor of more popular ones and missing commonsense travel decisions—but nonetheless gave her a few wonderful suggestions and pretty good guidance overall (Yeğinsu, 2024).

Therefore, since AI advice like human advice is a mix of good or bad, it is sometimes beneficial and sometimes disadvantageous to be influenced by it. So now we consider what causes advice to be influential and how does the influence of advice differ when it comes from an AI?

7.2 Antecedents to Influence

The key machine penalty outcome for advice is the amount of influence: *Are you influenced to actually follow a recommendation?* But before an AI or human can influence you, you have to receive their advice and think it's worth following. In more theoretical terms, influence is the primary outcome of the machine penalty for advice, but antecedents in the process of influence include both selecting an advisor and evaluating the advice itself. If there is evidence for the machine penalty in these antecedents, then we might not be surprised to find a machine penalty for the influence of advice itself.

7.2.1 Choosing an Advisor

Selecting an advisor is precursor to—and technically outside the scope of (*Similar Outcomes Condition*)—the machine penalty, as there is no outcome yet to evaluate. But in another way, it is at the heart of the machine penalty, because that selection is fully caused by the identity of the advisor, and therefore doesn't depend on the content of the advice—such as whether it's good or bad. Therefore, we'll learn a lot about how people are influenced by how they choose an advisor.

One study experimentally manipulated whether participants had to choose their advisor first before being assigned a task, or whether they were assigned the task and initially selected a response, but then got to choose an advisor. When the task was unassigned, about 75% of participants chose a human advisor over either a computer or robot, suggesting a precursor penalty. When the task was assigned and initially answered—and thus actual output—however, it depended on the task type. The majority of participants chose the human advisor for a social task but chose the computer or robot advisor for an analytical task (Hertz & Wiese, 2019).

For a similar difference in subjectivity, people believed humans were better suited for making predictions in the more subjective and moral domain of politics, and AI were better suited for making predictions in the more mathematical domain of economics. Yet when it came to actual influence, people were equally influenced by human or AI advice from both political and economic predictions (Himmelstein & Budescu, 2023). Likewise, people approved more of a human advisor compared to an AI advisor providing them with advice on legal cases, yet they were equally influenced by both (Aydin & Malle, 2024).

Finally, personalized decisions can also alter advisor selection. People preferred a human over an AI for providing personalized health and financial advice (Larkin et al., 2022), but only when the human is an expert in that area (Zhang et al., 2021) and is responsible for the advice they provide (Gazit et al., 2023).

7.2.2 Evaluating the Quality of the Advice

After selecting an advisor—or in cases where you received advice from multiple advisors—you then may discount the quality of the advice based on who or what it comes from.

Usually judging the advice as better or worse is related to following it, but it's also an important intermediate step in the process. For example, identical radiology advice coming from an AI instead of a human was perceived as lower quality when judged by radiology experts (Gaube et al., 2021). One way the perceived quality of AI advice is improved is actually by couching it as incorporating input from human experts (Zhang et al., 2024).

Even when advice suggests the same action, the rationale for it and who or what is providing that rationale can affect how we appraise it. Advice that was couched as a logical argument—why to drink enough water—was evaluated better when coming from AI than humans, whereas advice that seemed more like experienced-based tips—how to drink enough water—was perceived as better when coming from a human (Ahn et al., 2021).

Overall, we tend to see the machine penalty for both selecting an advisor and rating the quality of the advice, especially if the decision or advice itself is subjective, personalized, or social. In cases of more objective or logical decisions and advice, we often assume machines will be superior to humans.

7.3 When We Follow the Advice

Of course, the central measure of advice is whether we are influenced enough to follow it. In some cases we follow advice similarly from AI and humans (Himmelstein & Budescu, 2023), in other cases, we are more influenced by AI—such as the original study in algorithm appreciation (Logg et al., 2019)—and in other cases, we are most influenced by humans (Commerford et al., 2022). Let's consider what accounts for these different findings.

First, advice influence differs for subjective, hedonistic tasks that rely on human emotion and experience compared to more objective, utilitarian tasks that rely on quantification and calculation. There is a machine advantage on influence for AI recommendations over human recommendations in analytic tasks like estimation tasks (Mesbah et al., 2021), yet even some quantitative tasks like stock market predictions will show a machine penalty if statistical forecasts are compared to human experts (Önkal et al., 2009).

Research also shows we adhere more to an AI's recommendations on a range of less analytical, but still objective, tasks such as identifying a time period from a painting, a city from a photo taken of its landscape, or the income level of an individual based on their demographics. Yet we are only influenced more by humans on the most subjective, experiential tasks like identifying sarcasm from Reddit snippets (Vodrahalli et al., 2022). Likewise, people preference following the advice from humans on purchasing products for more hedonistic uses, but this disappears and often turns into a machine advantage for more objective or utilitarian uses of those products (Longoni & Cian, 2020; Wien & Peluso, 2021; Xie et al., 2022).

Second, influence also differs based on the essence, mind, or behavioral capacities of the AI such as whether they can learn and adapt. For example, estimating the amount of something—an objective task—was more influential when coming from an AI than a human, supporting the machine advantage. Yet when shown identical past performances for the human and AI, the human was then more influential than the AI, supporting the machine penalty. Finally, there was neither a machine penalty nor an advantage when the AI and humans both displayed learning through continuously improved performances (Berger et al., 2021).

If AI are unable to learn and improve, then bad advice at one point in time is likely to set low expectations for their future behavior. In one study, participants received advice that differed in its quality over time about hospital management tasks either from a human or computer. When the advice was good, people followed it regardless of who or what gave it. But after receiving poorer advice, people were then less likely to follow the subsequent good advice if it was from the computer system (Prahl & Van Swol, 2017). Conversely, when receiving advice on a personal or social issue, people initially preferred it from humans over ChatGPT. But in cases where they first saw high-quality advice from ChatGPT on other issues, their preference for its recommendation increased (Böhm et al., 2023).

A third difference in influence is the expertise of the AI or human advisor and the expertise of the one receiving the advice. While expertise increases the influence of both human and AI recommendations, it often gives an additional advantage to humans, especially for more hedonistic and subjective evaluations (Önkal et al., 2009; Xie et al., 2022). For example, while people trust novice human financial advisors more than AI advisors, they do not expect them to do better or want to hire them. Yet, expert human financial advisors are trusted more, expected to do better, and hired more than AI advisors (Zhang et al., 2021). Expertise can also be demonstrated through detailed advice information, such as people improving their accuracy when an AI advisor provides its level of certainty for each of its recommendations (Subramanian et al., 2023).

Those with some knowledge in a domain—potentially experts themselves—often realize the difficulty of providing quality advice and are especially suspicious of whether an AI can produce such advice. For example, accounting specialists who have some level of expertise were influenced less by AI, compared to human experts, for a complex auditing estimate (Commerford et al., 2022). Likewise, radiology experts discounted advice from AI compared to other human radiologists, whereas medical experts not trained in radiology did not distinguish between the advice (Gaube et al., 2021). On a binary basketball betting task, people balanced their own expertise with that of the AI advisor, but failed to increase their reliance on the AI when it had increased expertise (Elder et al., 2024)

A final way advice influence can differ relates to the personalization required for the advice. In many cases AI can actually provide high-quality personalized advice. People often find personalized AI advice just as good as human advice when they don't know who or what is providing it. Yet when the advisor is revealed, they display the machine penalty and are less influenced. For example, about three-fourths of participants chose a human advisor over ChatGPT for advice on a social or personal issue, while those who didn't know the identity of the advisor didn't distinguish between the recommendations of either (Böhm et al., 2023).

The importance of personalization and personal advice increases when it deals with moral, health, or life issues, usually exacerbating the machine penalty. We are less likely to comply with medical advice when it comes from an AI compared to a human physician (Larkin et al., 2022; Soellner & Koenigstorfer, 2021). The machine penalty for medical decisions is partly based on the people perceiving medical advice as needing to be unique and personalized to them. That means those who see themselves as more unique penalize the AI more, whereas it also means personalized AI medical advice is accepted as much as human medical advice (Longoni et al., 2019).

The influence penalty is additionally attributable to a related penalty for the responsibility of the medical advisor. When humans are not held responsible for the medical advice they provide, people's preference for their advice drops to the same level as their preference for AI's advice (Gazit et al., 2023).

7.4 The Machine Penalty for Advice

Overall, an influence machine penalty for advice is related to several of the machine penalty causes. AI or humans identified as experts lead us to expect better advice and then follow or be more influenced by it. We also expect the behavior of AI to be more stable, involving less adaptation and learning. So, when we observe their previous behavior, we see it as less malleable than that of humans and more likely to indicate their future behavior. This expectation can lead to a machine penalty if we've observed mistakes and poor recommendations but can lead to increased influence if we've experienced high-quality advice from AI.

Related to our assumptions about AI's mind and essence, we experience a strong machine penalty for subjective, experiential, and hedonistic advice. Yet there is similarly strong evidence for a machine advantage for advice clearly dealing with objective and utilitarian tasks. We're less likely to trust a machine's recommendation on which handmade chess set is most beautiful, but we know its suggestion of which move to make next is probably the best option. Similar to subjectivity, personalization and morality tap into a different human–machine essence, which makes us prefer humans to advise us more in personal decisions even when an AI could be potentially more helpful.

In summary, whether there is a machine penalty or advantage for giving advice is highly contingent on the causes at play, both human–AI comparisons and the situation.

7.5 Consequences: Reduced Versus Enhanced Performance

Diminishing advice from a machine can lead to tragedy—as in Air France Flight 447—or lead to averting tragedy—as in Stanislav Petrov's decision. Less dramatically, it can also lead to reduced or enhanced performance, depending on the situation. In situations when we diminish advice just because it comes from a machine, we often lose out on enhancing our performance or decision for several reasons.

First, AI advice is much easier to produce than human advice in most cases. While some general advice, like ratings of products and movies, is quite easy to provide, advice in specific situations and for specific individuals requires time from a human who is knowledgeable of that domain or person.

Second, AI advice is often high-quality even in domains we might not expect, like personalized plans, on social, moral, or health issues, or even with more subjective and experience-based decisions. Based on the machine penalty's situational causes and evidence presented, these are the exact situations in which the penalty is most likely and therefore where we are most likely to miss out on quality recommendations. Even if we believe a human should make the final decision on issues of personal health or ethical decisions, discounting the AI's advice—not its decision—only serves to ignore a potential source of insight.

That leads to the third, which is that AI advice is often part of more complex human–AI interactions such as human–AI teams and hybrid human–AI decision-making. This can lead to enhanced performance, for example, when an AI advisor for a StarCraft II strategy leads to enhanced performance on not only the current game, but on future games with no AI advisor (Tanksley et al., 2022).

Yet it can also lead to reduced performance. One research study showed that radiologists typically do not make better diagnoses with AI advice based on medical images, even though the AI, on average, outperforms human radiologists. This puzzling finding is explained by the fact that radiologists ignore or discount the advice from AI, especially when it conflicts with their views (Agarwal et al., 2023; Bresnick, 2023). Since they consider themselves experts and radiologists trust AI advice less than human expert advice (Gaube et al., 2021), they believe—or at least want to believe—that they are right. Therefore, they ignore the AI's suggestions.

Consequently, the lack of AI influence due to the radiologists' need for control and autonomy leads to worse medical diagnoses than the human–AI team could produce together (Reverberi et al., 2022). In extreme cases, fully penalizing a machine's advice can lead to tragedy, such as ignoring the clear warnings on Air France Flight 447.

We've also seen cases where a machine advantage is well-justified, and depending on the AI advice leads to enhanced performance. People often show the machine advantage for recommendations in quantitative, objective, and computational domains because they rightly understand AI to excel in such areas. To most efficiently win at Scrabble, it makes sense to take the computer's hint over your family member's suggestion. On average—unless your family member is a master player—following the Scrabble AI's advice is the best move, since it can compute the optimal decision in that situation.

But when advice is bad, a machine advantage produces negative consequences and a machine penalty is not only appropriate, but the best response. A machine penalty reduces our being influenced by bad advice.

AI, based on how they are trained, programmed, and learn, are known to reflect human biases (Morewedge et al., 2023; O'Neil, 2016), have specific idiosyncratic preferences and personalities, and can hallucinate, which is giving advice that is completely fabricated. Even simple malfunctions can generate completely wrong information and recommendations, like the missiles reported to Lt. Colonel Petrov.

Therefore, our performance for following such advice can be very problematic, especially if we blindly give the AI a full machine advantage, such as letting our GPS lead us into danger. It can also be dangerous if the advice deals with human morality, where large language models like ChatGPT can provide people with inconsistent ethical advice (Krügel et al., 2023). In cases where the advice is about other people, this can be especially damaging if we use advice that is systematically biased or unfair. In the next chapter, we consider a fairness machine penalty when AI and humans evaluate people, even when those evaluations are the same and aren't biased.

References

Agarwal, N., Moehring, A., Rajpurkar, P., & Salz, T. (2023). *Combining human expertise with artificial intelligence: Experimental evidence from radiology* (Working Paper 31422). National Bureau of Economic Research.

Ahn, J., Kim, J., & Sung, Y. (2021). AI-powered recommendations: The roles of perceived similarity and psychological distance on persuasion. *International Journal of Advertising, 40*(8), 1366–1384. https://doi.org/10.1080/02650487.2021.1982529

Associated Press. (2023). Man 'encouraged' by AI chatbot 'girlfriend' to kill Queen Elizabeth II receives jail sentence. *Euronews*. https://www.euronews.com/next/2023/10/06/man-encouraged-by-an-ai-chatbot-to-assassinate-queen-elizabeth-ii-receives-9-year-prison-s

Aydin, Z., & Malle, B. F. (2024). Dissociated responses to AI: Persuasive but not trustworthy? In L. K. Samuelson, S. L. Frank, M. Toneva, A. Mackey, & E. Hazeltine (Eds.), *Proceedings of the 46th annual conference of the cognitive science society* (Vol. 46, pp. 4410–4418).

Berger, B., Adam, M., Rühr, A., & Benlian, A. (2021). Watch me improve – Algorithm aversion and demonstrating the ability to learn. *Business & Information Systems Engineering, 63*(1), 55–68. https://doi.org/10.1007/s12599-020-00678-5

Böhm, R., Jörling, M., Reiter, L., & Fuchs, C. (2023). People devalue generative AI's competence but not its advice in addressing societal and personal challenges. *Communications Psychology, 1*(1), 32. https://doi.org/10.1038/s44271-023-00032-x

Bresnick, J. (2023). Does AI matter if human clinicians ignore it? *DHInsights*. https://www.dhinsights.org/news/does-ai-matter-if-human-clinicians-ignore-it

Commerford, B. P., Dennis, S. A., Joe, J. R., & Ulla, J. W. (2022). Man versus machine: Complex estimates and auditor reliance on artificial intelligence. *Journal of Accounting Research, 60*(1), 171–201. https://doi.org/10.1111/1475-679X.12407

El Atillah, I. (2023). Man ends his life after an AI chatbot 'encouraged' him to sacrifice himself to stop climate change. *euronews.next*. https://www.euronews.com/next/2023/03/31/man-ends-his-life-after-an-ai-chatbot-encouraged-him-to-sacrifice-himself-to-stop-climate-

Elder, H., Canfield, C., Shank, D. B., Rieger, T., & Hines, C. (2024). Knowing when to pass: The effect of AI reliability in risky decision contexts. *Human Factors, 66*(2), 348–362. https://doi.org/10.1177/00187208221100691

Gaube, S., Suresh, H., Raue, M., Merritt, A., Berkowitz, S. J., Lermer, E., et al. (2021). Do as AI say: Susceptibility in deployment of clinical decision-aids. *NPJ Digital Medicine, 4*(1), 31. https://doi.org/10.1038/s41746-021-00385-9

Gazit, L., Arazy, O., & Hertz, U. (2023). Choosing between human and algorithmic advisors: The role of responsibility sharing. *Computers in Human Behavior: Artificial Humans, 1*(2), 100009. https://doi.org/10.1016/j.chbah.2023.100009

Hansen, L. (2015). 8 drivers who blindly followed their GPS into disaster. *The Week*. https://theweek.com/articles/464674/8-drivers-who-blindly-followed-gps-into-disaster

Harford, T. (2016). Crash: How computers are setting us up for disaster. *The Guardian*. https://www.theguardian.com/technology/2016/oct/11/crash-how-computers-are-setting-us-up-disaster

Hertz, N., & Wiese, E. (2019). Good advice is beyond all price, but what if it comes from a machine? *Journal of Experimental Psychology: Applied, 25*(3), 386. https://doi.org/10.1037/xap0000205

Himmelstein, M., & Budescu, D. V. (2023). Preference for human or algorithmic forecasting advice does not predict if and how it is used. *Journal of Behavioral Decision Making, 36*(1), e2285. https://doi.org/10.1002/bdm.2285

Hoffman, D. (1999). I had a funny feeling in my gut. *The Washington Post*. https://www.washingtonpost.com/wp-srv/inatl/longterm/coldwar/shatter021099b.htm

Krügel, S., Ostermaier, A., & Uhl, M. (2023). ChatGPT's inconsistent moral advice influences users' judgment. *Scientific Reports, 13*(1), 4569. https://doi.org/10.1038/s41598-023-31341-0

Larkin, C., Drummond Otten, C., & Árvai, J. (2022). Paging Dr. JARVIS! Will people accept advice from artificial intelligence for consequential risk management decisions? *Journal of Risk Research, 25*(4), 407–422. https://doi.org/10.1080/13669877.2021.1958047

Logg, J. M., Minson, J. A., & Moore, D. A. (2019). Algorithm appreciation: People prefer algorithmic to human judgment. *Organizational Behavior and Human Decision Processes, 151*, 90–103. https://doi.org/10.1016/j.obhdp.2018.12.005

Longoni, C., Bonezzi, A., & Morewedge, C. K. (2019). Resistance to medical artificial intelligence. *Journal of Consumer Research, 46*(4), 629–650. https://doi.org/10.1093/jcr/ucz013

Longoni, C., & Cian, L. (2020). Artificial intelligence in utilitarian vs. hedonic contexts: The "word-of-machine" effect. *Journal of Marketing, 86*(1), 91–108. https://doi.org/10.1177/0022242920957347

Mesbah, N., Tauchert, C., & Buxmann, P. (2021). Whose advice counts more–man or machine? An experimental investigation of AI-based advice utilization. 54th Hawaii International Conference on System Sciences, Hawaii.

Morewedge, C. K., Mullainathan, S., Naushan, H. F., Sunstein, C. R., Kleinberg, J., Raghavan, M., & Ludwig, J. O. (2023). Human bias in algorithm design. *Nature Human Behaviour, 7*(11), 1822–1824. https://doi.org/10.1038/s41562-023-01724-4

O'Neil, C. (2016). *Weapons of math destruction: How big data increases inequality and threatens democracy*. Broadway Books.

Oliver, N., Calvard, T., & Potočnik, K. (2017). The tragic crash of flight AF447 shows the unlikely but catastrophic consequences of automation. *Harvard Business Review*. https://hbr.org/2017/09/the-tragic-crash-of-flight-af447-shows-the-unlikely-but-catastrophic-consequences-of-automation

Önkal, D., Goodwin, P., Thomson, M., Gönül, S., & Pollock, A. (2009). The relative influence of advice from human experts and statistical methods on forecast adjustments. *Journal of Behavioral Decision Making, 22*(4), 390–409. https://doi.org/10.1002/bdm.637

Prahl, A., & Van Swol, L. (2017). Understanding algorithm aversion: When is advice from automation discounted? *Journal of Forecasting, 36*(6), 691–702. https://doi.org/10.1002/for.2464

Reverberi, C., Rigon, T., Solari, A., Hassan, C., Cherubini, P., GI Genius CADx Study Group, & Cherubini, A. (2022). Experimental evidence of effective human–AI collaboration in medical decision-making. *Scientific Reports, 12*(1), 14952. https://doi.org/10.1038/s41598-022-18751-2

Schaich Borg, J., Sinnott-Armstrong, W., & Conitzer, V. (2024). *Moral AI: And how we get there*. Random House.

Soellner, M., & Koenigstorfer, J. (2021). Compliance with medical recommendations depending on the use of artificial intelligence as a diagnostic method. *BMC Medical Informatics and Decision Making, 21*, 236. https://doi.org/10.1186/s12911-021-01596-6

Subramanian, H. V., Canfield, C., Shank, D. B., & Kinnison, M. (2023). Combining uncertainty information with AI recommendations supports calibration with domain knowledge. *Journal of Risk Research, 26*(10), 1137–1152. https://doi.org/10.1080/13669877.2023.2259406

Tanksley, D., Arnold, C., Wilson, A., Shank, D. B., & Wunsch, D. C. (2022). A robot advisor to improve computerized game play. *European Scientific Journal, 18*(4), 50–60. https://eujournal.org/index.php/esj/article/view/15134

Vodrahalli, K., Daneshjou, R., Gerstenberg, T., & Zou, J. (2022). Do humans trust advice more if it comes from ai? An analysis of human-ai interactions. In *Proceedings of the 2022 AAAI/ACM conference on AI, ethics, and society* (pp. 763–777). https://doi.org/10.1145/3514094.3534150

Wien, A. H., & Peluso, A. M. (2021). Influence of human versus AI recommenders: The roles of product type and cognitive processes. *Journal of Business Research, 137*, 13–27. https://doi.org/10.1016/j.jbusres.2021.08.016

Xie, Z., Yu, Y., Zhang, J., & Chen, M. (2022). The searching artificial intelligence: Consumers show less aversion to algorithm-recommended search product. *Psychology & Marketing, 39*(10), 1902–1919. https://doi.org/10.1002/mar.21706

Yeğinsu, C. (2024). My first trip to Norway, with A.I. as a guide. *New York Times*. https://www.nytimes.com/2024/06/26/travel/norway-artficial-intelligence-planners.html

Zhang, L., Pentina, I., & Fan, Y. (2021). Who do you choose? Comparing perceptions of human vs robo-advisor in the context of financial services. *Journal of Services Marketing, 35*(5), 634–646. https://doi.org/10.1108/JSM-05-2020-0162

Zhang, Y., Tuk, M. A., & Klesse, A.-K. (2024). Giving AI a human touch: Highlighting human input increases the perceived helpfulness of advice from AI coaches. *Journal of the Association for Consumer Research, 9*(3), 344–356. https://doi.org/10.1086/730710

8

Evaluating People

8.1 Is It Fair?

Consider how you would feel if you were denied bail or given a harsher sentence, and then you found out it was based on a racially biased AI.

COMPAS software was designed to evaluate people—criminals specifically—as to their likelihood of committing another crime. The AI software uses a proprietary algorithm with data from 137 questions answered by defendants themselves or pulled from their criminal records. The questions included work background, information about social and family relationships, and residential history. It evaluates them by producing a risk score for recidivism, which is the risk of committing another crime (Angwin et al., 2016). We can imagine how useful a quantitative analysis tool could be if it helped judges better decide on sentencing and bail, which are heavily dependent on determining someone's risk of committing another crime.

When people are evaluated, our first concern is often whether it's a good and valid evaluation. In terms of the quality of its evaluation, COMPAS did okay, but not great. Its predictions were slightly better than chance with 61% of its predictions correctly assessing whether a

D. B. Shank, *The Machine Penalty*, https://doi.org/10.1007/978-3-031-86061-4_8

defendant would reoffend in the next two years (Angwin et al., 2016). However, it's arguable whether this is good enough since its predictions weren't any better than the predictions made by random people with no criminal justice experience (Dressel & Farid, 2018).

Usually for an AI to be routinely used to effectively evaluate people, it needs at least adequate, if not really good, predictions. But for the moment let's assume that 61% correct with 39% incorrect is acceptable in this domain—it's pretty difficult to determine if someone is going to commit a crime. Then our next and fundamentally more important concern should be whether that evaluation is fair, and this is doubly a concern for AI for a couple of reasons.

First, since humans program, select training data, and train AI and therefore can change them, we should only use them to evaluate others if they are both good and fair. With humans, sometimes we just have to take the best evaluators we can get. Second, AI as computer programs can be scaled up in a way that human evaluations can't be scaled up. If COMPAS is good *and* fair, then it could be implemented in courts all over the country and even the world, whereas a human judge—whether fair or unfair—can only take on so many cases.

Unfortunately, COMPAS was shown to be biased based on race according to a ProPublica analysis, even though race is not one of the 137 questions used as input (Angwin et al., 2016). Just because these questions didn't include race doesn't mean the data and the model are colorblind and bias-free. It's possible for other variables to be proxies for race—for example, if you know someone's neighborhood, you have a decent chance of correctly predicting their race. If you—or AI—know their neighborhood and 136 other variables including social connections and work history, you have an even better time predicting their race. And if these variables are also used to predict criminality, the model could be racially biased.

It also means the questions might be statistically good questions, but still inappropriate. Social connections and work history, for example, might be associated with criminality and risk for recidivism, but do we really want the criminal justice system sentencing people differently based on their jobs?

Thus, ProPublica decided to audit the COMPAS system. Their analysis didn't compare the average risk level of Whites and Blacks, because depending on the population, it's possible those average risk levels could be different but still be accurate (Humerick, 2020). That is, Whites and Blacks—at least the sample for their dataset—might collectively have different criminal backgrounds and thus different levels of risk.

Since the algorithm makes errors 39% of the time, ProPublica determined it would be racially biased if that error rate was systematically different for different races. And that's what their analysis found: COMPAS incorrectly labeled more Blacks than Whites as *high risk* of reoffending and incorrectly labeled more Whites than Blacks as *low risk* of reoffending.

One type of error is labeling defendants as *high risk*, but then finding out they don't go on to reoffend. Specifically, 23.5% of Whites, but 44.9% of Blacks, who were labeled *high risk* didn't reoffend—an error rate of Blacks nearly twice that of Whites. The other kind of error is labeling someone as *low risk* but then finding that they go on to commit another crime. ProPublica found that 28.0% of Blacks labeled as low risk reoffended, but 47.7% of Whites labeled as low risk went on to reoffend (Angwin et al., 2016).

So, both of these errors were biased against Blacks, advising judges who use the program that, inaccurately, a greater proportion of Blacks than Whites should be locked up. Whether or not we think 61% overall accuracy is good enough, COMPAS was systemically unfair across racial groups (Angwin et al., 2016) with additional analysis showing it was also systematically biased against Hispanics (Hamilton, 2019). Because of this racial bias, the COMPAS data has become an infamous case, with the dataset often used to audit AI algorithms for whether they'll be fair and unbiased. Yet, while it violates some types of fairness, such as those based on race, it was fair according to other fairness metrics (Guszcza et al., 2018; Humerick, 2020).

This example shows the importance of fairness when evaluating people using AI.

8.2 Bias in Evaluating People

AI aren't only used to evaluate criminals but all sorts of people for all sorts of reasons, and unfortunately, they often produce unfair, biased outcomes in those domains too (O'Neil, 2016). This is often because AI learns from human data and behavior, which includes biases against others (Morewedge et al., 2023; O'Neil, 2016).

AI can be biased in evaluating people when they apply for loans (Hale, 2021), housing (Raub, 2018; Rosen et al., 2021), and credit cards (Vigdor, 2019). Amazon scrapped a hiring AI they had designed themselves when they found out it was systematically biased against hiring women (Dastin, 2018). AI tools have shown bias when they're used to evaluate poor people's need for housing and other resources (Eubanks, 2018), for predicting the maltreatment of children (Eiermann, 2024), for predicting areas of higher crime and for targeting likely criminals (Eubanks, 2018; Ferguson, 2017). AI are used to evaluate students and teachers, often with little recourse when they receive poor and biased evaluations (O'Neil, 2016).

Of course, AI, like human evaluators, can be either fair or unfair. If an evaluator—whether AI or human—is truly fair and accurate, then it would be advantageous to use (Miller, 2018). Whereas if it is not, then it can produce poor or biased evaluations. However, since AI can be programmed and scaled up, and they often can't be appealed to directly, the consequences for a bad or unfair AI-based evaluation are often more severe (Eubanks, 2018; O'Neil, 2016; Wachter-Boettcher, 2017).

In this situation, a fairness machine penalty for evaluating people can mean caution toward AI evaluations, which is beneficial before adopting COMPAS or other biased software. Yet, the machine penalty could also manifest as diminishing outcomes of good and fair evaluations just because they are from AI. In this case, at best, we'd be missing out on the ability of AI to help our organizations and society make better decisions, but at worst we might be favoring biased humans over less biased AI.

8.3 Being Evaluated

Organizations are one of the primary places where people are routinely evaluated. They have many reasons to want to make fair and unbiased decisions, from believing it's the ethical way to treat people to, more pragmatically, avoiding lawsuits. We perceive all sorts of personnel decisions made within organizations to be less fair when made by AI, compared to human managers, especially by those who are being evaluated.

Evidence shows a fairness penalty against AI for evaluations and decisions related to promotion, termination (Newman et al., 2020), training (Bankins et al., 2022), employee evaluation (Bankins et al., 2022; Lee, 2018), recruitment, and hiring (Bankins et al., 2022; Lacroux & Martin-Lacroux, 2022; Lavanchy et al., 2023; Lee, 2018). Aside from being perceived as less fair, AI evaluations and decisions are also trusted less, perceived as less appropriate (Bankins et al., 2022), and produce more negative emotions (Lee, 2018), including the feeling of being dehumanized for those people who are evaluated by an AI (Jago & Laurin, 2017).

In a systematic literature review of AI in the workplace, Langer and Landers conclude that "most of the reviewed studies found that people perceive automated decisions as less fair than human decisions across personnel selection, performance evaluation, and scheduling scenarios" (2021, p. 10).

The few exceptions that don't support the machine penalty typically involve outside perceivers who are not being evaluated themselves and the general use of AI for evaluating people without a specific evaluation case. For example, university students reported that using AI evaluation for admissions of new students would be more fair than human evaluations to use in general across all applicants (Marcinkowski et al., 2020). In another study, job applicants considered the evaluation by an AI or human of their asynchronous interview as equally fair, but notably there was no actual interaction or evaluation provided (Suen et al., 2019).

We also penalize AI for decisions that relate to evaluating people who are not inside the organization doing the evaluation. We prefer policing decisions such as evaluating whether to stop someone being made by humans over AI. However, when the evaluations turn out to be correct, we are more likely to accept algorithmic evaluations as being similar to human evaluations in policing (Hobson et al., 2021).

One of the reasons the penalty occurs involves perceivers' assumptions that the human evaluators can consider more qualitative factors than the AI (Newman et al., 2020). Due to this assumption, interviewees may provide shorter responses and express themselves less when evaluated by an AI instead of a human (Langer et al., 2020).

This fairness machine penalty can be reduced, but not eliminated, when applicants receive a positive outcome or when the AI is known to take their unique characteristics into account (Lavanchy et al., 2023). The penalty may also be reduced if the machine feels more like a human through anthropomorphic features. For example, people who were negatively evaluated by an anthropomorphic robot felt like it treated them worse than people who received the same negative evaluation from a less anthropomorphic robot, although this was not compared to feedback by a human (Yam et al., 2022).

8.3.1 Being Monitored

Inside or outside of organizations, many of the evaluations of individuals are one-time, discrete evaluations such as hiring, firing, or policing. While one-time evaluations like hiring are important, ongoing evaluations like monitoring can introduce new concerns about privacy and autonomy, and ultimately fairness. While people may not like discrete evaluations, monitoring may be worse as it can feel invasive, and this might be different for AI compared to humans.

In one study, people perceived they had less autonomy, performed worse, and criticized and resisted more when they were surveilled by an AI compared to a human (Schlund & Zitek, 2021). Yet in another study, people report being less judged and having more autonomy when tracked by an AI compared to a human (Raveendhran & Fast, 2021).

These contrary results seem to stem from the implementation of slightly different types of monitoring: tracking versus surveillance. In

Schlund and Zitek's study, the surveillance was more proximate—happening in a real-time activity, recalling a time when it happened, or imagining it as a call center worker—and always involved a negative evaluative judgment by the AI or human (Schlund & Zitek, 2021).

In contrast, in Raveendhran and Fast's study, the tracking was more distal—choosing what type of tracking would occur in a future lab study, evaluating between job offers that implemented different types of tracking, or imagining tracking in a white-collar company. Additionally, their tracking was not as evaluative but gave personalized feedback for improvement by either a data analysis algorithm or a human analyst (Raveendhran & Fast, 2021).

Therefore, when viewed as more distant and beneficial, we prefer AI to monitor us. Whereas when it comes to negative feedback from direct monitoring, AI seems more intrusive, awakening a penalty against them.

Fascinatingly, both studies found that the reduction of autonomy was a primary mechanism that led to people's preferences for either human or AI. And both studies showed people were always less accepting of a monitoring process for evaluation purposes, rather than feedback or record-keeping (Raveendhran & Fast, 2021; Schlund & Zitek, 2021). Thus, the machine penalty, or machine advantage, for monitoring situations seems to be highly dependent on how controllable and personal it is, two of the situational causes of the penalty.

8.3.2 Being Managed

Different from but related to being evaluated or monitored, AI can also be used for managing people's tasks. Human managers often have multiple aspects to their jobs including evaluating, monitoring, and managing their subordinates. But aspects of managing such as assigning tasks and scheduling can be separated from the evaluations themselves, and often these tasks can also be relegated to AI.

In a field study of factory workers' task assignments, the workers favored AI picking their assignment over human managers, showing a machine advantage. However, the researchers discovered that one of the major reasons was because they were not being personally evaluated and they thought the AI would be more equitable overall for the quantitative distribution of tasks (Bai et al., 2021). Yet, while AI might be more equitable, they are seen as less benevolent and empathetic than human managers. This leads to a machine penalty for trusting the AI for management which requires higher levels of empathy (Li & Bitterly, 2024). Other research—which previously showed a machine penalty in fairness for evaluation—found no human–AI differences in trust or fairness for non-evaluative management decisions such as work assignments and scheduling (Lee, 2018).

Therefore, AI used for quantitative management tasks may be more preferred, whereas those used for management tasks that require empathy may be less preferred. Of course, even for quantitative management tasks, this doesn't mean there are no downsides to being managed by an AI. We perceive employees—ourselves or others—who are managed by an AI as having lower status and less complex jobs compared to those who are managed by a human (Jago et al., 2024).

8.4 Reacting to Unfair Outcomes

Finally, sometimes the issue isn't whether one thinks an AI or human is fairer in evaluating—or monitoring or managing—people, but rather in our reaction to unfair evaluations when they occur. Like Amazon's gender-biased hiring system and COMPAS's racially biased recidivism predictions, AI—like humans—can have systemic biases and make unfair decisions. This bias can come from biased data, unbalanced training, poor programming, or interactive learning. Regardless of its source, if we judge unfair outcomes differently when they are produced by AI compared to humans, then any response we make to condemn or correct this bias may be tempered by these different judgments.

For hiring recommendations specifically, people who observed an unfair decision against a minority or female candidate rated it as less discriminatory when from an AI compared to a human (Jago & Laurin, 2022). Similarly, we diminish the seriousness of errors in legal evaluations of defendants in court cases when they are made by an AI compared to the same errors by a human judge (Parlangeli et al., 2023). In contrast, there is not a clear difference between AI and humans in morality when the same decisions are fair toward minorities, although the humans are held more responsible (Hidalgo et al., 2021).

Overall, we do tend to perceive biased, unfair, or discriminatory evaluations and decisions as less wrong, and find the perpetrator less responsible and blameworthy when caused by AI instead of a human (Shank & DeSanti, 2018; Shank, DeSanti, et al., 2019). People are less morally outraged by AI that discriminate compared to humans who do the same, leading to less blame for this discrimination to their organization (Bigman et al., 2023).

These lower levels of blame and anger are based on the essentialist belief that AI can't be prejudiced or at least can't be as prejudiced as humans. When perception of prejudicial beliefs is taken into account—comparing AI and humans with the same level of perceived prejudicial beliefs—the diminished morality and blame disappears (Bigman et al., 2023).

Interestingly, this belief that machines can't be prejudiced or can't be prejudiced like humans, in some cases, leads us to prefer machines (Jago & Laurin, 2022). When people are expecting discrimination, they prefer to be evaluated by AI, precisely for the reasons they generally prefer humans: that AI will not treat them as humans with specific, yet irrelevant, qualities (Jago & Laurin, 2017) and that they won't be prejudiced toward them (Bigman et al., 2023; Jago & Laurin, 2022).

This process can also interact with larger social bias. When told about inequality in medical outcomes, such as racial disparities in hospitals, people's preferences shifted away from human evaluators toward AI evaluators (Bigman et al., 2021). People's underlying assumption here is that racial disparities in hospitals are more reflective of the people who work in those hospitals, not the algorithmic evaluators they use.

Finally, we are highly motivated to not recognize our own biases. In one research project, participants didn't perceive different levels of racial or gender bias when the same unfair outcomes were produced by other people or by algorithms trained on those other people's biased data. That is, there was no machine penalty in general against recognizing bias. Yet, people saw their own biased outcomes as much less biased compared to an algorithm trained on their own biased data (Celiktutan et al., 2024).

Here, a machine penalty is less about reacting to unfair outcomes, but about our self-serving interest in appearing unbiased.

In some cases, it may be arguable whether an outcome is fair or not because it taps into other human values. For example, a judge or AI either forgave or gave a full sentence to a man who robbed a store to steal food for his sick wife, pitting justice motivations against mercy motivations. In this case the human judge was held more responsible than the AI for the forgiveness, but they were held equally responsible for the full sentence. Interestingly, the human judge is seen as more moral than the AI for both outcomes (Hidalgo et al., 2021).

8.5 The Machine Penalty for Evaluating People

Overall, I might conclude that there are two different penalties related to evaluating people. First, there is a fairness penalty, where we diminish the fairness of AI in making general evaluations of people (Sect. 8.3). Second, there is an unfairness penalty, where we diminish the unfairness of AI—perceiving them as less unfair, which, of course, also means fairer—when an AI or human has made a clearly biased evaluation of people (Sect. 8.4).

For the first, we tend to diminish the fairness of AI even when they may produce valid and accurate evaluations of people. There is quite a bit of evidence for a fairness penalty for both organizational and non-organizational decisions that evaluate people, whereas there is less evidence for a machine penalty when the AI is only monitoring or managing without evaluating. Both those being evaluated and third-party observers tend to display a machine penalty for evaluation, although those who are evaluated often produce stronger reactions.

When it comes to understanding the complexities and uniqueness of humans and their performance and roles, we tend to believe AI aren't as sophisticated as people at making those evaluations. This means we generally prefer—and perceive it as fairer—for humans, over AI, to evaluate us, but if we suspect prejudice then an AI that doesn't consider our "uniqueness" is preferred.

Yet, for the second penalty, when biased outcomes that we see as unfair are produced, we generally judge AI as less unfair, including less wrong, being less angered by them, and attributing less responsibility and blame to them. While there is evidence for both diminishing fairness and unfairness in these different situations, it's not evident whether there is a tipping point between biased and unbiased outcomes, or if this is based on framing an outcome as being fair or unfair. That's an important area for future research.

The penalties for evaluating people depend on several machine penalty causes such as the subjective nature of the evaluation, our essential beliefs, and mind perception of AI being less capable of prejudice compared to humans. When we are the ones being evaluated, it's also dependent on our control of the situation, the personal nature of the evaluation, and its general importance.

8.6 Consequences: Checked Versus Unchecked Bias

When AI is used to evaluate people, the key consequence of a fairness machine penalty involves the spread versus the restriction of bias. Depending on the situation, the machine penalty can keep bias in check or let it continue unchecked.

Biased programs, like COMPAS, need to be discontinued, and new algorithms need to be programmed with safeguards and trained with debiased training data (Morewedge et al., 2023). AI should also undergo audits for fairness (Guszcza et al., 2018), and important decisions made by AI should include explainable AI and an appeal process for those who believe they have been judged unfairly (Eubanks, 2018; O'Neil, 2016). When AI includes explanations of their decisions, people are better able to detect and mitigate bias (Hou et al., 2024), but that presumes accurate explanations from the AI. Likewise, organizations should set up processes to monitor AI systems and solicit feedback from the employees who are directly influenced by those systems in order to avoid issues both from the AI and how they're perceived.

Bias is an ongoing issue as AI systems are becoming more complex, are trained on more massive datasets, and can learn from real-world interactions with data, people, and environment. While the bias of the COMPAS program was widely known after being detailed in the ProPublica report, many cases are, unfortunately, not that obvious.

Here, the machine penalty is like a double-edged sword. The machine penalty suggests we won't want to use or adopt AI programs like COMPAS for evaluating people, especially if we know that an AI's evaluation is unfair. In contrast, if that program is adopted and actually unfair, the machine penalty makes us less likely to perceive unfairness in that AI system compared to a human. If, hypothetically, COMPAS was able to be compared to a human judge that produced the same biased recidivism scoring, we would presumably see the judge as more intentionally prejudiced and thus rate the judge's scores as even more unfair than COMPAS. That higher level of unfairness for the human, not the AI, would be more likely to prod us to take action. Therefore, the penalty provides a check in general for AI programs for evaluation but can leave currently occurring AI bias unchecked.

Yet not all AI programs have strong biases, and those designed carefully and trained with appropriate datasets can produce high-quality and fair outcomes. There is a real possibility of a fairness machine penalty leading us to curb the use and influence of these good AI evaluators (Miller, 2018). This is a more knee-jerk version of the penalty where we simply consider AI as less appropriate for personal or somewhat subjective decisions, not because it's actually worse in evaluating people.

Finally, when unfairness or bias does occur, a penalty toward AI can lead to an increased responsibility gap (Matthias, 2004)—where AI can increase bias, without anyone or anything having increased responsibility for causing that bias. In this case, when we fail to blame an AI or blame an AI less, it doesn't always mean that we find another entity to share the blame. Because AI is programmed by humans, trained on data, produced by a company, and used by another person or organization, this leads to a problem of many hands (Thompson, 1980), wherein many entities could be blamed, but often none are. This lack of responsibility can also be exploited by those humans who control the AI evaluators, effectively shielding them from liability (Rosen et al., 2021) while allowing them to act unethically (Gratch & Fast, 2022).

While dramatic stories about AI unfairness like COMPAS or Amazon's hiring algorithm often make the headlines, this opposite concern is that when AI are unfair toward people, our reactions will be so tepid that this bias will continue unchallenged. This leaves open the possibility that the machine penalty can lead to ongoing unfairness, which harms the people who are evaluated. In this next chapter we consider harm, including actual physical harm, and what happens when AI is designed—but sometimes fails—to keep people safe.

References

Angwin, J., Larson, J., Mattu, S., & Kirchner, L. (2016). Machine Bias. *ProPublica*. https://www.propublica.org/article/machine-bias-risk-assessments-in-criminal-sentencing

Bai, B., Dai, H., Zhang, D., Zhang, F., & Hu, H. (2021). The impacts of algorithmic work assignment on fairness perceptions and productivity. *Academy of Management Proceedings, 2021*(1), 12335. https://doi.org/10.5465/AMBPP.2021.175

Bankins, S., Formosa, P., Griep, Y., & Richards, D. (2022). AI decision making with dignity? Contrasting workers' justice perceptions of human and AI decision making in a human resource management context. *Information Systems Frontiers, 24*(3), 857–875.

Bigman, Y. E., Wilson, D., Arnestad, M. N., Waytz, A., & Gray, K. (2023). Algorithmic discrimination causes less moral outrage than human discrimination. *Journal of Experimental Psychology: General, 152*(1), 4–27. https://doi.org/10.1037/xge0001250

Bigman, Y. E., Yam, K. C., Marciano, D., Reynolds, S. J., & Gray, K. (2021). Threat of racial and economic inequality increases preference for algorithm decision-making. *Computers in Human Behavior, 122*, 106859. https://doi.org/10.1016/j.chb.2021.106859

Celiktutan, B., Cadario, R., & Morewedge, C. K. (2024). People see more of their biases in algorithms. *Proceedings of the National Academy of Sciences, 121*(16), e2317602121. https://doi.org/10.1073/pnas.2317602121

Dastin, J. (2018). Amazon scraps secret AI recruiting tool that showed bias against women. *Reuters*. https://www.reuters.com/article/idUSKCN1MK0AG/

Dressel, J., & Farid, H. (2018). The accuracy, fairness, and limits of predicting recidivism. *Science Advances, 4*(1), eaao5580. https://doi.org/10.1126/sciadv.aao5580

Eiermann, M. (2024). Algorithmic risk scoring and welfare state contact among US children. *Sociological Science, 11*, 707–742. https://doi.org/10.15195/v11.a26

Eubanks, V. (2018). *Automating inequality: How high-tech tools profile, police, and punish the poor*. St. Martin's Press.

Ferguson, A. G. (2017). *Rise of big data policing, The*. New York University Press.

Gratch, J., & Fast, N. J. (2022). The power to harm: AI assistants pave the way to unethical behavior. *Current Opinion in Psychology, 47*, 101382. https://doi.org/10.1016/j.copsyc.2022.101382

Guszcza, J., Rahwan, I., Bible, W., Cebrian, M., & Katyal, V. (2018). Why we need to audit algorithms. *Harvard Business Review*. https://hbr.org/2018/11/why-we-need-to-audit-algorithms

Hale, K. (2021). A.I. bias caused 80% of black mortgage applicants to be denied. *Forbes*. https://www.forbes.com/sites/korihale/2021/09/02/ai-bias-caused-80-of-black-mortgage-applicants-to-be-denied/

Hamilton, M. (2019). The biased algorithm: Evidence of disparate impact on Hispanics. *American Criminal Law Review, 56*, 1553–1577.

Hidalgo, C. A., Orghian, D., Canals, J. A., De Almeida, F., & Martin, N. (2021). *How humans judge machines*. MIT Press.

Hobson, Z., Yesberg, J. A., Bradford, B., & Jackson, J. (2021). Artificial fairness? Trust in algorithmic police decision-making. *Journal of Experimental Criminology, 19*, 1–25. https://doi.org/10.1007/s11292-021-09484-9

Hou, T.-Y., Tseng, Y.-C., & Yuan, C. W. T. (2024). Is this AI sexist? The effects of a biased AI's anthropomorphic appearance and explainability on users' bias perceptions and trust. *International Journal of Information Management, 76*, 102775. https://doi.org/10.1016/j.ijinfomgt.2024.102775

Humerick, J. (2020). Reprogramming fairness: Affirmative action in algorithmic criminal sentencing. *Columbia Human Rights Law Review, 4*, 213–245. https://hrlr.law.columbia.edu/hrlr-online/reprogramming-fairness-affirmative-action-in-algorithmic-criminal-sentencing/

Jago, A. S., & Laurin, K. (2017). *Algorithms and dehumanization in hiring*. Academy of Management Atlanta, GA.

Jago, A. S., & Laurin, K. (2022). Assumptions about algorithms' capacity for discrimination. *Personality and Social Psychology Bulletin, 48*(4), 582–595. https://doi.org/10.1177/01461672211016187
Jago, A. S., Raveendhran, R., Fast, N., & Gratch, J. (2024). Algorithmic management diminishes status: An unintended consequence of using machines to perform social roles. *Journal of Experimental Social Psychology, 110*, 104553. https://doi.org/10.1016/j.jesp.2023.104553
Lacroux, A., & Martin-Lacroux, C. (2022). Should I trust the artificial intelligence to recruit? Recruiters' perceptions and behavior when faced with algorithm-based recommendation systems during resume screening. *Frontiers in Psychology, 13*, 895997. https://doi.org/10.3389/fpsyg.2022.895997
Langer, M., König, C. J., & Hemsing, V. (2020). Is anybody listening? The impact of automatically evaluated job interviews on impression management and applicant reactions. *Journal of Managerial Psychology, 35*(4), 271–284. https://doi.org/10.1108/JMP-03-2019-0156
Langer, M., & Landers, R. N. (2021). The future of artificial intelligence at work: A review on effects of decision automation and augmentation on workers targeted by algorithms and third-party observers. *Computers in Human Behavior, 123*, 106878. https://doi.org/10.1016/j.chb.2021.106878
Lavanchy, M., Reichert, P., Narayanan, J., & Savani, K. (2023). Applicants' fairness perceptions of algorithm-driven hiring procedures. *Journal of Business Ethics, 188*(1), 125–150. https://doi.org/10.1007/s10551-022-05320-w
Lee, M. K. (2018). Understanding perception of algorithmic decisions: Fairness, trust, and emotion in response to algorithmic management. *Big Data & Society, 5*(1), 2053951718756684. https://doi.org/10.1177/2053951718756684
Li, M., & Bitterly, T. B. (2024). How perceived lack of benevolence harms trust of artificial intelligence management. *Journal of Applied Psychology*. https://doi.org/10.1037/apl0001200
Marcinkowski, F., Kieslich, K., Starke, C., & Lünich, M. (2020). *Implications of AI (un-)fairness in higher education admissions: The effects of perceived AI (un-)fairness on exit, voice and organizational reputation*. Proceedings of the 2020 conference on fairness, accountability, and transparency, Barcelona, Spain. https://doi.org/10.1145/3351095.3372867
Matthias, A. (2004). The responsibility gap: Ascribing responsibility for the actions of learning automata. *Ethics and Information Technology, 6*, 175–183. https://doi.org/10.1007/s10676-004-3422-1

Miller, A. P. (2018). Want less-biased decisions? Use Algorithms. *Harvard Business Review*. https://hbr.org/2018/07/want-less-biased-decisions-use-algorithms
Morewedge, C. K., Mullainathan, S., Naushan, H. F., Sunstein, C. R., Kleinberg, J., Raghavan, M., & Ludwig, J. O. (2023). Human bias in algorithm design. *Nature Human Behaviour, 7*(11), 1822–1824. https://doi.org/10.1038/s41562-023-01724-4
Newman, D. T., Fast, N. J., & Harmon, D. J. (2020). When eliminating bias isn't fair: Algorithmic reductionism and procedural justice in human resource decisions. *Organizational Behavior and Human Decision Processes, 160*, 149–167. https://doi.org/10.1016/j.obhdp.2020.03.008
O'Neil, C. (2016). *Weapons of math destruction: How big data increases inequality and threatens democracy*. Broadway Books.
Parlangeli, O., Currò, F., Palmitesta, P., & Guidi, S. (2023). Asymmetries in the moral judgements for human decision-makers and Artificial Intelligence Systems (AI) delegated to make legal decisions. In *Proceedings of the European Conference on Cognitive Ergonomics 2023* (pp. 1–4, Article 16). https://doi.org/10.1145/3605655.3605676
Raub, M. (2018). Bots, bias and big data: Artificial intelligence, algorithmic bias and disparate impact liability in hiring practices. *Arkansas Law Review, 71*(2), 529–570.
Raveendhran, R., & Fast, N. J. (2021). Humans judge, algorithms nudge: The psychology of behavior tracking acceptance. *Organizational Behavior and Human Decision Processes, 164*, 11–26. https://doi.org/10.1016/j.obhdp.2021.01.001
Rosen, E., Garboden, P. M., & Cossyleon, J. E. (2021). Racial discrimination in housing: How landlords use algorithms and home visits to screen tenants. *American Sociological Review, 86*(5), 787–822. https://doi.org/10.1177/00031224211029618
Schlund, R., & Zitek, E. (2021). Who's my manager? Surveillance by AI leads to perceived privacy invasion and resistance practices. In *Academy of Management Proceedings*. https://doi.org/10.5465/AMBPP.2021.11451abstract
Shank, D. B., & DeSanti, A. (2018). Attributions of morality and mind to artificial intelligence after real-world moral violations. *Computers in Human Behavior, 86*, 401–411. https://doi.org/10.1016/j.chb.2018.05.014
Shank, D. B., DeSanti, A., & Maninger, T. (2019). When are artificial intelligence versus human agents faulted for wrongdoing? Moral attributions after

individual and joint decisions. *Information, Communication, and Society, 22*(5), 648–663. https://doi.org/10.1080/1369118X.2019.1568515
Suen, H.-Y., Chen, M. Y.-C., & Lu, S.-H. (2019). Does the use of synchrony and artificial intelligence in video interviews affect interview ratings and applicant attitudes? *Computers in Human Behavior, 98*, 93–101. https://doi.org/10.1016/j.chb.2019.04.012
Thompson, D. F. (1980). Moral responsibility of public officials: The problem of many hands. *American Political Science Review, 74*(4), 905–916. https://doi.org/10.2307/1954312
Vigdor, N. (2019). Apple card investigated after gender discrimination complaints. *The New York Times*. https://www.nytimes.com/2019/11/10/business/Apple-credit-card-investigation.html
Wachter-Boettcher, S. (2017). *Technically wrong: Sexist apps, biased algorithms, and other threats of toxic tech*. WW Norton & Company.
Yam, K. C., Goh, E.-Y., Fehr, R., Lee, R., Soh, H., & Gray, K. (2022). When your boss is a robot: Workers are more spiteful to robot supervisors that seem more human. *Journal of Experimental Social Psychology, 102*, 104360. https://doi.org/10.1016/j.jesp.2022.104360

9

Causing Harm

9.1 Who's to Blame?

In March 2018 in Tempe, Arizona, a Volvo SUV Uber crashed into Elaine Herzberg while she crossed a road with her bicycle, killing her instantly (Wu, 2023). Like other harm from car accidents, we want to know who is to blame, but unlike other accidents, this was the first deadly automobile accident involving an autonomous vehicle. In addition to that unique status, there were a number of other factors that complicated and made this accident a case study in the attribution of blame. Let's consider the line-up of those who could potentially be blamed.

Herzberg herself crossed without looking and illegally, as there was no pedestrian crossing. She was wearing dark clothes with no reflectors, and she may have been affected by methamphetamine, which was found in her blood when she died (Schaich Borg et al., 2024). Thus, there are several reasons to hold her at least somewhat responsible for her own death.

Yet the autonomous vehicle failed to detect and identify Herzberg as a pedestrian that needed to be avoided (Schaich Borg et al., 2024). This seems like it should be a basic safety feature built into every autonomous vehicle, and so such a failure may be the worst kind of technological error.

D. B. Shank, *The Machine Penalty*, https://doi.org/10.1007/978-3-031-86061-4_9

Unfortunately, it might have been more than just a one-time mistake. People had been complaining about the increased number of accidents with Uber's autonomous vehicles, suggesting that there were underlying safety concerns with them that had not adequately been addressed (National Transportation Safety Board, 2019; Schaich Borg et al., 2024).

Some of the issues that affected this collision were that Uber's AI system would wait to brake until it could fully identify an object that was in its way, and it was not designed to specifically account for jaywalking pedestrians. This might suggest that not only the technology, but its designers at Uber, Volvo, or other companies that contributed to the technology might also share some blame (Schaich Borg et al., 2024).

Of course, programmers and technology designers create software and hardware based on what a company's leadership decides should be developed and deployed, including what risks were acceptable (Vanek, 2023). The National Transportation Safety Board concluded that Uber engaged in risky practices that overall disregard human safety (2019). Legally, there seemed to be some potential for blame here, as Herzberg's family members reached an undisclosed settlement with Uber (Wu, 2023).

Knowing these safety concerns, California had previously enacted more regulation and threatened legal actions against Uber, leading Uber to discontinue its autonomous vehicle tests in California cities. Arizona had the opposite response, inviting Uber to test their autonomous vehicles in their cities, including Tempe where the crash that killed Herzberg occurred. Therefore, we might put some blame on Arizona's state government for its lax rules and safety protocols, while inviting its streets to be a testbed for somewhat risky technologies (Schaich Borg et al., 2024). The courts disagreed that the government could be legally blamed, dismissing the lawsuit by Herzberg's family against the city of Tempe (Wu, 2023).

While all of these entities could be—and are—blamed to different degrees, when the dust finally settled, Rafaela Vasquez—who I've yet to mention—probably received most of the blame. Vasquez was the human driver known as the "safety driver" of this autonomous Uber Volvo (Hawkins, 2023). She was supposed to provide oversight and take control of the vehicle, if needed, in the case of an emergency, even though generally the vehicle was expected to fully drive itself.

If Vasquez had been watching the road, she would have had just enough time to brake the vehicle and avoid hitting Herzberg. Yet she was looking away and distracted, streaming an episode of *The Voice* on her phone (Coppola & Frank, 2018). That might seem like a clear case of negligence, and therefore blame. But as with everything in this case, it was not that simple. Vasquez actually had two phones: a personal phone and a company phone. The company phone displayed messages related to her next Uber assignments, and Uber required her to check those messages as they came in. Therefore, she maintains she was following the company's procedures by looking away to check her work phone, and that she was not watching, but only listening to, *The Voice* using her personal phone (Schaich Borg et al., 2024). While originally charged with negligent homicide, Vasquez eventually pleaded guilty to a lesser charge of endangerment and received three years of probation (Vanek, 2023; Wu, 2023).

Jana Schaich Borg and colleagues, who have an excellent extended discussion of blame and responsibility in this situation, overview it like this:

> Who or what was responsible for Herzberg's death? Herzberg? Vasquez? The car? Uber? Safety managers at Uber? Engineers at Uber? Arizona government officials who allowed Uber to test their cars in Tempe? All of them? None of them? (2024, p. 137)

When an accident like this occurs and people are harmed, one of the first big questions we ask is, *Who's to blame*? Not only do we as observers ask that but also court proceedings, news coverage, and social media posts often revolve around this question of blame and related questions of whether it was intentional, who was responsible, and who is liable?

Not only can blame impact insurance payments, compensation, fines, and even prison sentences, but also, especially when it results in culpability and punishment, it can help bring about emotional closure for victims and their families. At a societal level, this process provides some moral solace as we want offenders to receive appropriate punishment for the harm they caused. Societally, it can also lead to changes in designs, training, and laws.

Ultimately, blame is a central outcome when an AI or human causes harm. No one wants and no AI is programmed to have an accident or let a patient suffer and die, but nevertheless, accidents occur, and medical issues are not all dealt with perfectly. Even in military operations targeting enemy combatants, the questions of blame usually revolve around mistakes, or the harm done to unintentional targets like allied soldiers and innocent civilians. Therefore, the presence of a moral situation does not in and of itself bring about blame, but only when harm is caused do we look for someone or something to blame.

9.2 Harm as Morality

A situation that can potentially cause harm is a moral situation and decisions made within it are moral decisions because they deal with human harm. While some moral situations, decisions, and outcomes involve moral categories other than harm (Graham et al., 2013), harm itself has been argued to be the central type of moral violation (Gray et al., 2014; Schein & Gray, 2018) across different cultures and political orientations (Gray & Pratt, 2025). Specifically, we tend to think moral violations are wrong insomuch as they harm someone. Likewise, when we see a behavior as harmful, then we begin to label it as immoral (Gray et al., 2014; Schein & Gray, 2018).

It is critical to frame AI harm and blame under the umbrella of morality for three reasons.

First, computers are not only *social* actors (Reeves & Nass, 1996), but AI and machines are increasingly considered *moral* actors (Gamez et al., 2020). Based on that, there is a moral psychology of artificial intelligence (Bonnefon et al., 2024; Ladak et al., 2024), which applies theories and research of morality including harm to AI as wrongdoers.

Second, morality research also gives insights into the process of blame (Shaver, 1985), built on a long tradition of attributions of responsibility (Kelley, 1972; Weiner, 1986). Specifically, Bertram Malle and colleagues mapped out our cognitive process of blame, identifying the agent's causality, intentionality, reasons, obligation to prevent the event, and capacity to prevent the event as the key factors that lead to blame, or different degrees of blame (Malle et al., 2014). These factors interacting with other key human–AI comparisons change who or what we blame across situations (Bigman et al., 2019; Monroe et al., 2014).

Third, decisions that can involve moral outcomes are regulated, meaning it's not permissible for everyone. Gun ownership and use require permits and using one in the military requires training and a commission. Likewise, we require a medical license to make crucial health decisions, and various licenses to operate heavy machinery, fly airplanes, or drive cars. Because these domains where harm can be caused are regulated, before we get to blame, we need to ask whether machines should actually make moral decisions.

9.3 Machines Making Moral Decisions

Who do we want to make moral decisions, especially moral decisions that are potentially harmful or harm-mitigating? Research shows that we prefer humans, not AI, to make these moral decisions, indicating a machine penalty for moral permissibility (Bigman & Gray, 2018; Shank & DeSanti, 2018; Shank, DeSanti, & Maninger, 2019).

Specifically, we find it more permissible for humans compared to AI to make life-and-death decisions in areas as diverse as driving, surgery, and military attacks. This penalty is somewhat explained by differences in the perceived mind of AI compared to humans (Bigman & Gray, 2018). We expect humans to have moral minds, and, consequently, to effectively make these decisions.

And that's true even if the AI are equally as capable (Bigman & Gray, 2018). In one study, people read a complex ethical dilemma scenario involving when to shoot down a terrorist drone with a bomb which varied both the chance and the number of innocent people that could be killed. While people perceived the expert AI system as more capable and reliable than a human expert, they still placed more moral trust and responsibility in the human expert. That is, they saw it as morally more permissible for the human to make the actual life-and-death decision (Tolmeijer et al., 2022).

We also find it less permissible for AI to make moral decisions when it could have a negative effect on us directly. We generally prefer AI drivers to save the most people, but we quickly change our preferences to favor the vehicle's passengers over pedestrians if we are in the vehicle or purchasing it for ourselves (Bonnefon et al., 2016; Gill, 2020; Liu & Liu, 2021). Due to our desire for controllability, we prefer vehicles that allow us to switch to a manual mode to take control, even if the AI has superior driving and even if manual mode vehicles cost more (Arnestad et al., 2024).

Part of the reason for a machine penalty for moral permissibility may be that machine morality must be programmed or learned, and therefore AI can be given a specific set of moral principles or rules to follow. But people often disagree on moral choices, so if an AI is going to make moral choices whose ethical values should AI be based on? We sometimes believe that others' ethical values are not correct, and this can lead us to avoid, challenge, or simply disagree with those other people. For AI, we don't want those preferences codified and scaled-up into systems that we use, especially given there is the capacity to program them with morals different from ours.

In a series of studies, people preferred utilitarian decisions for autonomous vehicles—that is to protect the largest number of people possible (Awad et al., 2018; Bonnefon et al., 2016; Li et al., 2016). Yet utilitarian choices are not always as simple as they seem. What if there are the same number of people? And does it matter who those people are? And what about whether they are passengers or pedestrians, or obeying the traffic laws or not?

One option to address these moral questions would be to program AI to emulate the aggregate moral preferences across humans. This could still be biased, but it would be biased in the same way as humans in the aggregate are biased. In the moral machine experiments, millions of people from countries all over the world were given trolley-style moral dilemmas with varied characteristics of those who could be killed. Trolley dilemmas are when one has to decide to kill one person or group of people in order to save another person or group. In the moral machine experiments, the decision was for a vehicle to swerve to avoid pedestrians, either killing specific people in the car or on the street (Awad et al., 2018).

This data shows, overall, people prefer sparing humans over pets, pedestrians over passengers, women over men, healthy people over unhealthy, the young over the old, those obeying the law over those who are not, and higher-status people over lower-status people. However, these "worldwide" preferences are averages and differ individually and culturally (Awad et al., 2018). Also, they don't consider the important factors of probability and severity of harm, which would impact a real-world trolley dilemma for driving (Krügel & Uhl, 2022). Therefore, while some guidelines exist for quantifying moral choices, there is far from a universally accepted set of moral values for machines to use (Bigman & Gray, 2020).

9.4 Blame for Harm

Moving to blame itself, let's first consider the paradigmatic case of blame for harm from a vehicle accident. In some cases, research shows machine penalty where the AI is blamed less than the human (Awad et al., 2020; Beckers et al., 2022; Li et al., 2016; Mu & Karasawa, 2024; Stojilovi*f*á et al., 2024), yet in other cases show the opposite, wherein the AI is blamed more for the accident (Hidalgo et al., 2021; Hong et al., 2020; Liu et al., 2019; Liu & Du, 2022; Young & Monroe, 2019; Zhang, Tuk, & Klesse, 2024). Yet in other cases still, AI and human drivers receive nearly equal amounts of blame (Hong et al., 2021).

Considering other contexts of AI harm, people also blame AI less than humans for building decisions that lead to a bridge collapse, for police ambushes that harm innocent people (Hidalgo et al., 2021), for military decisions that unintentionally kill civilians (Tolmeijer et al., 2022), and for airplane crashes (Gómez-Sánchez et al., 2023). For risky evacuation decisions that save or fail to save people from a tsunami, fire, or hurricane, humans were blamed or held more responsible for their decision than AI, in spite of whether the outcome was harmful or harm-mitigating (Hidalgo et al., 2021). In one case, where a police officer or AI vision system shoots an innocent person who happens to match the description of a terrorist, the human and AI system received similar blame (Hidalgo et al., 2021).

Across a range of norm-violating harmful behaviors, including physically harmful ones like pulling a dog's tail to hitting a stranger in the face, humans receive more blame than AI-powered robots (Monroe et al., 2014). Similarly, across emotionally harmful behavior like telling someone they are overweight and physically painful behaviors like jabbing an animal with a stick, people perceive humans as more blameworthy than AI (Maninger & Shank, 2022). Finally, the same is true for more relational emotional harm, like an elder caregiver switching care away from a patient that has an emotional connection to them to another patient (Hidalgo et al., 2021).

In a trolley dilemma where a mining train will hit and kill one or multiple coal miners, people blame AI more than humans when they don't save the most lives, but humans more than AI when they do (Malle et al., 2015). Likewise, across a series of trolley problem variants, Malle and colleagues found that human and AI decision-makers often receive similar levels of blame. The exception is in the cases where inaction doesn't save the most lives, in which case AI are blamed more than humans (Malle et al., 2025).

Overall, there is quite a bit of evidence for a harm machine penalty, except for the paradigmatic case of self-driving vehicles which has very mixed support. To understand the mixed support, we need to consider the specifics of those studies. Within those we'll see how easily blame can be shifted by details about the specific situation, the different harms caused, the AI's presentation, and whether the AI or human is making the decision or providing oversight. Let's consider each in turn.

9.4.1 Situational Factors

Some situations change the focus of the blame and who could logically be expected to have reacted in time to stave off the harm. Across different accident or near accident scenarios, researchers found that AI were blamed more than human drivers for most incidents, because the situations causing the accident were unexpected—such as a person stepping out from behind a parked tractor-trailer truck. When the situation was more expected—such as a person stepping out from behind a parked bus—blame increased for both humans and AI, but more so for human drivers, meaning they were blamed most. AI is expected to be more cautious and react faster to unexpected surprise situations (Zhang, Tuk, & Klesse, 2024).

Additionally, details revealed about the motivations and situation alter blame. People who tended to initially blame human drivers more than AI for an accident changed to similar levels of blame after learning additional details about the driver's intentionality, and the foreseeability and preventability of the accident. The similar levels of blame were across both positive and negative information such as the driver intentionally or unintentionally hitting the car, suggesting that situational knowledge in general helps equalize AI and human blame. However, knowledge of preventability and intentionality also raised the overall level of blame, including the number of respondents who fully blamed the AI or human (Stojilovi*f*á et al., 2024).

In contrast, when motivation was given for running a red light—transporting a sick woman to the hospital—the perceived intentionality increased more for humans than for AI. This higher level of intentionality accounted for increased blame of the human, compared to the AI, for causing an accident (Mu & Karasawa, 2024).

9.4.2 Differing Levels of Harm

Another difference in a penalty for blame is the type of harm.

In multiple studies that found AI were blamed more than humans for causing an accident while driving, blame increased for all drivers more for accidents than for near accidents (Zhang, Tuk, & Klesse, 2024) and for causing death over non-fatal injuries (Hong et al., 2020). In these cases, AI blame increased more than human blame (Hong et al., 2020; Zhang, Tuk, & Klesse, 2024). Similarly, people blamed AI more than human drivers for a crash not directly due to its harm, but to feeling worse about that crash—more negative emotions led to this difference in blame (Liu et al., 2019; Liu & Du, 2022).

In moral dilemma situations like the trolley dilemma, someone is always harmed, and the decision is only who will be harmed. In this case, people may blame an AI more than a human for any decision they make (Young & Monroe, 2019) or at least if they didn't make the utilitarian one to save the most lives (Malle et al., 2015). Here the increase— or machine advantage—in blame is due to a machine penalty for morality. Because we believe that machines can't and shouldn't make moral decisions, when they do make moral decisions that lead to harm—though unavoidable in a moral dilemma—we blame them more for that harmful outcome.

Considering other moral behaviors can highlight how blame differs specifically for harm. Exploring helping behavior that prevents harm, researchers showed that AI drivers were praised more than their human counterparts for driving a man to the hospital when he suffered a medical emergency (Hong et al., 2021). Comparing harm to other moral violations, my student and I found that across immoral behaviors, we blame AI less than humans and perceive their behaviors as less morally wrong. Interestingly, people were more likely to misidentify the type of immoral behaviors when done by an AI and less likely to blame the AI compared to a human. But harmful behaviors acted differently: people overwhelmingly identified them correctly as a harm and thought they more directly deserved blame, regardless of whether an AI or human was the perpetrator (Maninger & Shank, 2022).

9.4.3 AI Presentation

Specific AI presentation and assumed capacities can also affect the blame-game for harm.

In the trolley dilemma, when people perceive the AI as having a more humanlike mind, the human–AI differences in this blame go away (Young & Monroe, 2019), because any entity perceived as having more mind is viewed as more responsible for harm (Nijssen et al., 2023). Automated vehicles that are anthropomorphized with voice, name, and gender are trusted more than those that are not, because they too are perceived as possessing a more humanlike mind (Waytz et al., 2014). Across research, anthropomorphism and related mind judgments are key to moral blame (Gray et al., 2012) including for robots and AI (Bigman et al., 2019; Gray & Wegner, 2012).

In another experiment, an AI-powered farm robot increased the risk of poisoning people from the irrigation runoff by deciding to use a novel fertilizer. When the robot was described as more sophisticated including anthropomorphic features and mental capacities, people blamed it more, even though they didn't find the risky, potentially harmful decision any more wrong (Kneer & Stuart, 2021).

People blamed autonomous AI weapons differently than human military drone pilots for the decision to either continue or abort a strike on terrorists when a civilian child could be injured. Research study participants blamed AI similarly for either the launch or abort decision, whereas humans were blamed more for aborting than launching. This meant AI were sometimes blamed more or less than humans, depending on the decision made. However, these differences were accounted for by expectations that the human pilot had to follow orders, whereas it was less clear if the AI should follow orders or make the optimal decision at the time. When the human–AI identity difference of being in a command structure was taken into account, blame was similar for humans and AI (Malle et al., 2019).

9.4.4 Position as Driver or Oversight

We are interested in whether we should blame a machine or human in situations similar to the accident that killed Elaine Herzberg, wherein a human driver has oversight of an autonomous vehicle with the automated AI system doing the actual driving. In cases like this, the comparison is often between humans and AI in different roles but in the same situation, as opposed to comparisons across situations.

In these cases, in general, blame tends to go to the primary driver (Mu & Karasawa, 2024) or to whomever or whatever makes an error or mistake (Awad et al., 2020). Yet when both humans and AI make mistakes, the human bears the brunt of the blame compared to the machine (Awad et al., 2020). Similarly, if a human providing oversight is distracted they also receive more blame than if they are not distracted (Beckers et al., 2022), yet level of automation—for example, fully, partial, and with or without override—has more impact on blame than the type of distraction (Wotton et al., 2022).

In a full comparison of oversight scenarios, human drivers received even more blame when they were monitored by another human compared to having AI oversight. And humans or AI giving oversight were blamed less if they were overseeing a human, not an AI. Finally, humans giving oversight were blamed more than AI giving oversight, indicating a machine penalty for oversight of driving (Mu & Karasawa, 2024).

As these suggest, humans can easily be attributed the blame for harm when in a position of oversight. Simply having the ability to shift to manual control or override the automation increases blame for the humans (Arnestad et al., 2024). Humans who drive themselves or override the AI both receive more blame for an accident and more praise for acting selflessly to avoid an accident than if the AI is driving (McManus & Rutchick, 2018). But even humans who have the option to override but don't, or humans who choose the AI's programming, also receive more blame or praise respectively than AI (McManus et al., 2024; McManus & Rutchick, 2018).

For humans, it's not surprising that blame for oversight is at the same level of a human driving themselves (Copp et al., 2023). However in other human–AI collaborations that lead to harm, humans who simply oversee AI receive less blame than humans who take action themselves or do so after advice from AI (Shank, DeSanti, & Maninger, 2019). Likewise, when a human taking an AI recommendation about mushrooms ends up poisoning their friend, blame is given to the actors most proximate to causing the harm (Leichtmann et al., 2024).

9.5 The Machine Penalty for Harm

The cases which show a machine advantage—blaming an AI more than a human for harm—are often explained by insights into a machine's behaviors and essence. Machines are—or should—be programmed to avoid accidents, minimize damage, and preserve lives. Additionally, we expect reactions from machines to be cautious, quick, and objective. Therefore, we sometimes hold them to higher expectations when they are used on the battlefield, in surgeries, or on the road. When harm occurs in these cases, blame follows.

Yet, there is substantial evidence for machine penalties, diminishing an outcome from an AI, for harm. We find it less permissible for AI to make moral decisions and don't always agree on what decisions they should make. Because of this, when harm is unavoidable—as in trolley-style moral dilemmas—we find AI more blameworthy for their decisions, especially if they don't maximize the number of lives saved. When unintended harm occurs, we blame machines less because we see them as less minded and intentional than humans. We also blame them less if there is potentially a human overseer to blame. As with Herzberg, even an imperfect AI system might not be blamed as much as an imperfect human.

Blame for harm is connected in some way to nearly all the machine penalty causes, including various human–AI comparisons of appearance, identity, behavior, mind and essence. And the situation itself is critical, not only harm being a moral outcome, but the intrinsic importance of human life. Adding to this important, personal, and controllable causation means extra scrutiny if you are the one letting an AI drive or make the decision about your medical operation.

9.6 Consequences: Eroded Versus Increased Safety

Depending on when we blame AI less for harm, we may be eroding or increasing safety.

Like other moral outcomes, a machine advantage for blame might be appropriate if the AI is meaningfully perpetuating harm. For the car that killed Herzberg, blaming the vehicle's AI could lead to the companies involved, like Uber, implementing decisions that lead to increased safety even if it's at the expense of other aspects, such as slower driving time or braking more often. Not surprisingly, research shows people blame the company and victims more when an autonomous vehicle causes accidents compared to a human driver (McManus et al., 2024; Stojilovifá et al., 2024). This suggests that companies should be acutely aware of the blame they receive for harm done by their AI.

Yet blaming AI more for harm may actually lead to eroded safety if that harm is not widespread or is due to factors beyond the purview of the situation. It's unlikely that autonomous vehicles will ever be completely accident-free, but that doesn't mean we won't reduce accidents substantially from what we experience with human drivers. This suggests a psychological blockade, wherein a machine penalty can help temper an outsized level of blame and rejection for safe technologies (Shariff et al., 2017). This will keep us from upending our progress toward increased safety, where we balance the benefits and potential harm of both humans and AI.

Like the fairness penalty in Chap. 8, a diminishing blame for AI has the potential to create a responsibility gap (Matthias, 2004; Thompson, 1980), wherein no one is blamed for harm that occurs from AI (Nyholm, 2018). Yet, instead of simply diminishing blame when AI is the cause, blame can be shifted to the companies that create and sell the AI systems (McManus et al., 2024). Additionally, there is a possibility people will use AI as proxies for immoral and harmful acts (Bonnefon et al., 2024), as people can more easily avoid blame and punishment when delegating to a machine compared to a human (Bigman et al., 2019; Feier et al., 2022).

Physical harm including death is usually more obvious and visceral than unfair algorithms, and so it is more likely to be investigated and challenged. Safety checks, audits, and beta-testing must be part of the solution to increased safety with AI. Yes, AI can cause harm, but too much blame curtails their potential to reduce it as well.

For a completely different type of AI outcome, I next consider AI producing art.

References

Arnestad, M. N., Meyers, S., Gray, K., & Bigman, Y. E. (2024). The existence of manual mode increases human blame for AI mistakes. *Cognition, 252*, 105931. https://doi.org/10.1016/j.cognition.2024.105931

Awad, E., Dsouza, S., Kim, R., Schulz, J., Henrich, J., Shariff, A., et al. (2018). The moral machine experiment. *Nature, 563*(7729), 59–64. https://doi.org/10.1038/s41586-018-0637-6

Awad, E., Levine, S., Kleiman-Weiner, M., Dsouza, S., Tenenbaum, J. B., Shariff, A., et al. (2020). Drivers are blamed more than their automated cars when both make mistakes. *Nature Human Behaviour, 4*, 134–143. https://doi.org/10.1038/s41562-019-0762-8

Beckers, N., Siebert, L. C., Bruijnes, M., Jonker, C., & Abbink, D. (2022). Drivers of partially automated vehicles are blamed for crashes that they cannot reasonably avoid. *Scientific Reports, 12*(1), 16193. https://doi.org/10.1038/s41598-022-19876-0

Bigman, Y. E., & Gray, K. (2018). People are averse to machines making moral decisions. *Cognition, 181*, 21–34. https://doi.org/10.1016/j.cognition.2018.08.003

Bigman, Y. E., & Gray, K. (2020). Life and death decisions of autonomous vehicles. *Nature, 579*, E1–E2. https://doi.org/10.1038/s41586-020-1987-4

Bigman, Y. E., Waytz, A., Alterovitz, R., & Gray, K. (2019). Holding robots responsible: The elements of machine morality. *Trends in Cognitive Sciences, 23*(5), 365–368. https://doi.org/10.1016/j.tics.2019.02.008

Bonnefon, J.-F., Rahwan, I., & Shariff, A. (2024). The moral psychology of Artificial Intelligence. *Annual Review of Psychology, 75*, 653–675. https://doi.org/10.1146/annurev-psych-030123-113559

Bonnefon, J.-F., Shariff, A., & Rahwan, I. (2016). The social dilemma of autonomous vehicles. *Science, 352*(6293), 1573–1576. https://doi.org/10.1126/science.aaf2654

Copp, C. J., Cabell, J. J., & Kemmelmeier, M. (2023). Plenty of blame to go around: Attributions of responsibility in a fatal autonomous vehicle accident. *Current Psychology, 42*, 6752–6767. https://doi.org/10.1007/s12144-021-01956-5

Coppola, C., & Frank, B. J. (2018). Uber driver was watching 'The Voice' moments before fatal Tempe crash. *Azcentral*. https://www.azcentral.com/story/news/local/tempe-breaking/2018/06/21/uber-self-driving-car-crash-tempe-police-elaine-herzberg/724344002/

Feier, T., Gogoll, J., & Uhl, M. (2022). Hiding behind machines: Artificial agents may help to evade punishment. *Science and Engineering Ethics, 28*(2), 19. https://doi.org/10.1007/s11948-022-00372-7

Gamez, P., Shank, D. B., Arnold, C., & North, M. (2020). Artificial virtue: The machine question and perceptions of moral character in artificial moral agents. *AI & Society, 35*, 795–809. https://doi.org/10.1007/s00146-020-00977-1

Gill, T. (2020). Blame it on the self-driving car: How autonomous vehicles can alter consumer morality. *Journal of Consumer Research, 47*(2), 272–291. https://doi.org/10.1093/jcr/ucaa018

Gómez-Sánchez, J., Gordo, C., Franklin, M., Fernandez-Basso, C., & Lagnado, D. (2023). Who is to blame? Responsibility attribution in AI systems vs human agents in the field of air crashes. In H. L. Larsen, M. J. Martin-Bautista, M. D. Ruiz, T. Andreasen, G. Bordogna, & G. De Tré (Eds.), *Flexible query answering systems* (pp. 256–264). Springer Nature Switzerland. https://doi.org/10.1007/978-3-031-42935-4_21

Graham, J., Haidt, J., Koleva, S., Motyl, M., Iyer, R., Wojcik, S. P., & Ditto, P. H. (2013). Moral foundations theory: The pragmatic validity of moral pluralism. In P. Devine & A. Plant (Eds.), *Advances in experimental social psychology* (Vol. 47, pp. 55–130). Academic Press. https://doi.org/10.1016/B978-0-12-407236-7.00002-4

Gray, K., & Pratt, S. (2025). Morality in our mind and across cultures and politics. *Annual Review of Psychology, 76*. https://doi.org/10.1146/annurev-psych-020924-124236

Gray, K., Schein, C., & Ward, A. F. (2014). The myth of harmless wrongs in moral cognition: Automatic dyadic completion from sin to suffering. *Journal of Experimental Psychology: General, 143*(4), 1600–1615. https://doi.org/10.1037/a0036149

Gray, K., & Wegner, D. M. (2012). Feeling robots and human zombies: Mind perception and the uncanny valley. *Cognition, 125*(1), 125–130. https://doi.org/10.1016/j.cognition.2012.06.007

Gray, K., Young, L., & Waytz, A. (2012). Mind perception is the essence of morality. *Psychological Inquiry, 23*(2), 101–124. https://doi.org/10.1080/1047840X.2012.651387

Hawkins, A. J. (2023). Uber driver in first-ever deadly self-driving crash pleads guilty. *The Verge*. https://www.theverge.com/2023/7/31/23814474/uber-self-driving-fatal-crash-safety-driver-plead-guilty

Hidalgo, C. A., Orghian, D., Canals, J. A., De Almeida, F., & Martin, N. (2021). *How humans judge machines*. MIT Press.

Hong, J.-W., Cruz, I., & Williams, D. (2021). AI, you can drive my car: How we evaluate human drivers vs. self-driving cars. *Computers in Human Behavior, 125*, 106944. https://doi.org/10.1016/j.chb.2021.106944

Hong, J.-W., Wang, Y., & Lanz, P. (2020). Why is artificial intelligence blamed more? Analysis of faulting artificial intelligence for self-driving car accidents in experimental settings. *International Journal of Human–Computer Interaction, 36*(18), 1768–1774. https://doi.org/10.1080/10447318.2020.1785693

Kelley, H. H. (1972). Causal schemata and attribution process. In E. E. Jones, D. E. Kanouse, H. H. Kelley, R. E. Nisbett, S. Valins, & B. Weiner (Eds.), *Attribution: Perceiving the causes of behavior* (pp. 151–174). General Learning Press.

Kneer, M., & Stuart, M. T. (2021). *Playing the Blame Game with Robots* Companion of the 2021 ACM/IEEE International Conference on Human-Robot Interaction, Boulder, CO, USA. https://doi.org/10.1145/3434074.3447202

Krügel, S., & Uhl, M. (2022). Autonomous vehicles and moral judgments under risk. *Transportation Research Part A: Policy and Practice, 155*, 1–10. https://doi.org/10.1016/j.tra.2021.10.016

Ladak, A., Loughnan, S., & Wilks, M. (2024). The moral psychology of artificial intelligence. *Current Directions in Psychological Science, 33*(1), 27–34. https://doi.org/10.1177/09637214231205866

Leichtmann, B., Hinterreiter, A., Humer, C., Ventura, A., Streit, M., & Mara, M. (2024). Moral reasoning in a digital age: Blaming artificial intelligence for incorrect high-risk decisions. *Current Psychology*. https://doi.org/10.1007/s12144-024-06658-2

Li, J., Zhao, X., Cho, M.-J., Ju, W., & Malle, B. F. (2016). *From trolley to autonomous vehicle: Perceptions of responsibility and moral norms in traffic accidents with self-driving cars.* (2016-01-0164). SAE Technical Paper.

Liu, P., & Du, Y. (2022). Blame attribution asymmetry in human–automation cooperation. *Risk Analysis, 42*(8), 1769–1783. https://doi.org/10.1111/risa.13674

Liu, P., Du, Y., & Xu, Z. (2019). Machines versus humans: People's biased responses to traffic accidents involving self-driving vehicles. *Accident Analysis & Prevention, 125*, 232–240. https://doi.org/10.1016/j.aap.2019.02.012

Liu, P., & Liu, J. (2021). Selfish or utilitarian automated vehicles? Deontological evaluation and public acceptance. *International Journal of Human–Computer Interaction, 37*(13), 1231–1242. https://doi.org/10.1080/10447318.2021.1876357

Malle, B. F., Guglielmo, S., & Monroe, A. E. (2014). A theory of blame. *Psychological Inquiry, 25*(2), 147–186. https://doi.org/10.1080/1047840X.2014.877340

Malle, B. F., Magar, S. T., & Scheutz, M. (2019). AI in the sky: How people morally evaluate human and machine decisions in a lethal strike dilemma. In I. A. Ferreira, J. S. Sequeira, G. S. Virk, E. E. Kadar, & O. Tokhi (Eds.), *Robots and well-being* (pp. 111–133). Springer. https://doi.org/10.1007/978-3-030-12524-0_11

Malle, B. F., Scheutz, M., Arnold, T., Voiklis, J., & Cusimano, C. (2015). Sacrifice one for the good of many?: People apply different moral norms to human and robot agents. In *Proceedings of the Tenth Annual ACM/IEEE International Conference on Human-Robot Interaction* (pp. 117–124). ACM. https://doi.org/10.1145/2696454.2696458

Malle, B. F., Scheutz, M., Cusimano, C., Voiklis, J., Komatsu, T., Thapa, S., & Aladia, S. (2025). People's judgments of humans and robots in a classic moral dilemma. *Cognition, 254*, 105958. https://doi.org/10.1016/j.cognition.2024.105958

Maninger, T., & Shank, D. B. (2022). Perceptions of violations by artificial and human actors across moral foundations. *Computers in Human Behavior Reports, 5*, 100154. https://doi.org/10.1016/j.chbr.2021.100154

Matthias, A. (2004). The responsibility gap: Ascribing responsibility for the actions of learning automata. *Ethics and Information Technology, 6*, 175–183. https://doi.org/10.1007/s10676-004-3422-1

McManus, R. M., Mesick, C. C., & Rutchick, A. M. (2024). Distributing blame among multiple entities when autonomous technologies cause harm. *Personality and Social Psychology Bulletin, 01461672241238303*. https://doi.org/10.1177/01461672241238303

McManus, R. M., & Rutchick, A. M. (2018). Autonomous vehicles and the attribution of moral responsibility. *Social Psychological and Personality Science, 10*(3), 345–352. https://doi.org/10.1177/1948550618755875

Monroe, A. E., Dillon, K. D., & Malle, B. F. (2014). Bringing free will down to Earth: People's psychological concept of free will and its role in moral judgment. *Consciousness and Cognition, 27*, 100–108. https://doi.org/10.1016/j.concog.2014.04.011

Mu, Y., & Karasawa, M. (2024). Blame attribution and intentionality perception of human versus robot drivers: Implications for judgments about autonomous vehicles in moral dilemma contexts. *Cogent Psychology, 11*(1), 2384298. https://doi.org/10.1080/23311908.2024.2384298

National Transportation Safety Board. (2019). Collision between vehicle controlled by developmental automated driving system and pedestrian tempe, Arizona (NTSB/HAR-19/03 PB2019-101402). *National Transportation Safety Board.* https://www.ntsb.gov/investigations/Pages/HWY18MH010.aspx

Nijssen, S. R., Müller, B. C., Bosse, T., & Paulus, M. (2023). Can you count on a calculator? The role of agency and affect in judgments of robots as moral agents. *Human–Computer Interaction, 38*(5–6), 400–416. https://doi.org/10.1080/07370024.2022.2080552

Nyholm, S. (2018). Attributing agency to automated systems: Reflections on human–robot collaborations and responsibility-loci. *Science and Engineering Ethics, 24*(4), 1201–1219. https://doi.org/10.1007/s11948-017-9943-x

Reeves, B., & Nass, C. (1996). *The media equation: How people treat computers, television, and new media like real people and places*. Cambridge University.

Schaich Borg, J., Sinnott-Armstrong, W., & Conitzer, V. (2024). *Moral AI: And how we get there*. Random House.

Schein, C., & Gray, K. (2018). The theory of dyadic morality: Reinventing moral judgment by redefining harm. *Personality and Social Psychology Review, 22*(1), 32–70. https://doi.org/10.1177/1088868317698288

Shank, D. B., & DeSanti, A. (2018). Attributions of morality and mind to artificial intelligence after real-world moral violations. *Computers in Human Behavior, 86*, 401–411. https://doi.org/10.1016/j.chb.2018.05.014

Shank, D. B., DeSanti, A., & Maninger, T. (2019). When are artificial intelligence versus human agents faulted for wrongdoing? Moral attributions after individual and joint decisions. *Information, Communication, and Society, 22*(5), 648–663. https://doi.org/10.1080/1369118X.2019.1568515

Shariff, A., Bonnefon, J.-F., & Rahwan, I. (2017). Psychological roadblocks to the adoption of self-driving vehicles. *Nature Human Behaviour, 1*, 694–696. https://doi.org/10.1038/s41562-017-0202-6

Shaver, K. G. (1985). *The attribution of blame: Causality, responsibility, and blameworthiness*. Springer-Verlag.

Stojilovi*f*á, D., Franklin, M., Malle, B. F., Fernandez-Basso, C., Awad, E., & Lagnado, D. (2024). Are autonomous vehicles blamed differently? In *Proceedings of the annual meeting of the cognitive science society* (Vol. 46). https://escholarship.org/uc/item/7q35p9b7

Thompson, D. F. (1980). Moral responsibility of public officials: The problem of many hands. *American Political Science Review, 74*(4), 905–916. https://doi.org/10.2307/1954312

Tolmeijer, S., Christen, M., Kandul, S., Kneer, M., & Bernstein, A. (2022). Capable but amoral? Comparing AI and human expert collaboration in ethical decision making. In *Proceedings of the 2022 CHI Conference on Human Factors in Computing Systems* (pp. Article 160). https://doi.org/10.1145/3491102.3517732

Vanek, C. (2023). Arizona driver in fatal autonomous Uber crash in 2018 pleads guilty, sentenced to probation. *azcentral*. https://www.azcentral.com/story/news/local/tempe/2023/07/28/rafaela-vasquez-pleads-guilty-in-in-fatal-uber-self-driving-crash-killed-pedestrian-elaine-herzberg/70488361007/

Waytz, A., Heafner, J., & Epley, N. (2014). The mind in the machine: Anthropomorphism increases trust in an autonomous vehicle. *Journal of Experimental Social Psychology, 52*, 113–117. https://doi.org/10.1016/j.jesp.2014.01.005

Weiner, B. (1986). *An attributional theory of motivation and emotion*. Springer-Verlag.

Wotton, M. E. L., Bennett, J. M., Modesto, O., Challinor, K. L., & Prabhakharan, P. (2022). Attention all 'drivers': You could be to blame, no matter your behaviour or the level of vehicle automation. *Transportation Research Part F: Traffic Psychology and Behaviour, 87*, 219–235. https://doi.org/10.1016/j.trf.2022.03.021

Wu, D. (2023). A self-driving Uber killed a woman. The backup driver has pleaded guilty. *The Washington Post*. https://www.washingtonpost.com/nation/2023/07/31/uber-self-driving-death-guilty/

Young, A. D., & Monroe, A. E. (2019). Autonomous morals: Inferences of mind predict acceptance of AI behavior in sacrificial moral dilemmas. *Journal of Experimental Social Psychology, 85*, 103870. https://doi.org/10.1016/j.jesp.2019.103870

Zhang, Y., Tuk, M. A., & Klesse, A.-K. (2024). Giving AI a human touch: Highlighting human input increases the perceived helpfulness of advice from AI coaches. *Journal of the Association for Consumer Research, 9*(3), 344–356. https://doi.org/10.1086/730710

10

Producing Art

10.1 Is It Valuable?

The value of art is a tricky thing. *The Portrait of Edmond de Belamy* "depicts a portly gentleman, possibly French and — to judge by his dark frockcoat and plain white collar — a man of the church" (Christies.com, 2018). But what sets this artwork apart from others is not just that it was created by an AI, but that in 2018, it was sold for $432,000, making it the most expensive AI-produced piece of art ever sold (Katatikarn, 2024). In contrast, the largest sum paid for human artwork topped $450 million when the *Salvator Mundi*, or "Savior of the World" by Leonardo da Vinci was auctioned in 2017 (Dimarco, 2024).

The comparison here might seem straightforward: top human art is 100 times more valuable than top AI art, at least given the caveat that AI art still seems in its infancy. But another lesson is that the value of art is related to the artist's identity: da Vinci versus the AI programmed by *Obvious*, a Paris-based art collective that used a generative adversarial network to create *The Portrait of Edmond de Belamy*. The value of the artist's identity is seen in the monetary difference between a forgery and

D. B. Shank, *The Machine Penalty*, https://doi.org/10.1007/978-3-031-86061-4_10

authentic artwork, even though the quality and aesthetics are similar to all but a few art experts. *Salvator Mundi* had been sold in 2005—12 years prior to the record-breaking sale—for only $1175, listed as "After Leonardo da Vinci" meaning that it was thought to be someone's imitation of da Vinci, not an actual da Vinci (Blostein et al., 2018; Doward, 2019).

But the authentic identity of the artist isn't the only aspect that can lead to value. In 1996, Thomas Taylor had recently graduated from art school and was commissioned to do his first professional piece—a book cover no less—for about $650. The pencil sketch and concentrated watercolors took him two days after some initial drafts. The piece itself is good, but not a masterpiece, and Mr. Taylor has had a successful career as a writer and illustrator but is not a famous artist. So, why did this $650 watercolor sell for $1.92 million in 2024, an unheard-of price for a book cover illustration? Because Mr. Taylor's illustration was the original artist's cover for *Harry Potter and the Philosopher's Stone* (Yoon, 2024). Again, the authenticity of this unique piece was critical, but also the identity—not the individual artist's identity, but the work's identity—as it connects to the beloved Harry Potter series.

When considering why *The Portrait of Edmond de Belamy* was sold for more money than any other piece of AI art, identity doesn't really explain it. However, authenticity partially does, since AI artwork is often considered inauthentic (Jago, 2019) and sometimes not even fully considered art (Horton Jr. et al., 2023; Mikalonytė & Kneer, 2022). *The Portrait of Edmond de Belamy* challenged AI art's inauthentic status as it was the first AI artwork to be sold at a major auction. While other AI art might be sold by major auctions, this piece has the unique status of being the first. And through the auction process, *The Portrait of Edmond de Belamy* was further legitimized as real authentic art. Specifically, it was sold by Christie's, the 250-year-old auction house that also sold the *Salvator Mundi* for $450 million, and it was in the same lot as other reputable pieces of art, like Andy Warhol screenprints, and art by Banksy, Jeff Koons, and Christo Javacheff (Kinsella, 2018). Its inclusion in this auction process meant that it had been vetted by others, which allowed it to reach a certain standard that previously was only allowed by human art.

This monetary value of high-end art demonstrated through auctions reflects AI art's legitimacy and value to society, including how much it can be later sold to someone else. In contrast, the more individualistic value of art often relates to aesthetics. That is, the question of an artwork's value is answered at a societal level by considering money, but at an individual level by considering aesthetics. For a lot of art, this aesthetic value is more important as we don't actually pay much money, if any, to easily access high-quality images, videos, music, books, and other writing.

One example of a well-liked song is *Heart on My Sleeve*, garnering over 11 million views in just a few days after its release in April 2023, which put it on track to break into the Billboard's top songs. That, of course, wouldn't have been too surprising, as it came out as a collaboration between Drake and The Weeknd, two Canadian music artists who have individually and jointly enjoyed the success of many hit songs. Yet, *Heart on My Sleeve* was an AI-generated song using their voices without their permission to create a near-perfect mimicry and therefore was taken down and not allowed to earn royalties. Nonetheless, while it didn't lead to profits, it was still highly popular (Coscarelli, 2023; Kelly, 2023).

The aesthetic value of *Heart on My Sleeve* is also tied in several ways to the artist's authentic identity. Musician Holly Herdon touches on the aspects of identity by asking, "Do we care what Drake really feels or is it enough to just hear a superficially intelligent rendering? For some people that will not be enough" (Coscarelli, 2023). That is, some people value the song for the music itself, and for others what matters is the identity of the artist and that artist's authentic feelings that went into it.

Universal Music Group, the label that represents Drake and The Weeknd, put identity and authenticity more starkly as a battle of two sides, "the side of the artists, fans, and human creative expression" versus "the side of deep fakes, fraud, and denying artists their due compensation" (Kelly, 2023). While the song wasn't authentically Drake and The Weeknd, it also wasn't legitimately produced and distributed, with websites removing and putting disclaimers due to copyright violations (Coscarelli, 2023; Kelly, 2023). Nonetheless, a lot of people still like it.

Taken altogether, when we consider art, we often consider if it is valuable, either monetarily or aesthetically. And this question of value is often but not always tied to issues of authenticity and identity. With these ideas established, we can now dive into whether and when we diminish art created by machines.

10.2 Diminishing the Value of AI Art

Aesthetic value relates to our personal view of whether we like, enjoy, or find something high-quality, and this is often what we care about in our consumption of art. We buy, listen to, display, watch, and are most engaged with the creations we like. Often this is the entire story: we only care if we like it and don't really care who or what made it. And since most of the time people can't distinguish between art created by an AI or a human (Oksanen et al., 2023), the artist's identity and authenticity are of secondary concern to aesthetics.

Thus, when art is identified as being produced by an AI or human, we might not care because this identity makes no difference to us. That's why there is sometimes no machine penalty for aesthetics: specifically for visual art (Gu & Li, 2022; Hong & Curran, 2019; Messingschlager & Appel, 2023), music (Tigre Moura & Maw, 2021; Zlatkov et al., 2023), and poetry (Chu & Liu, 2023; Hitsuwari et al., 2023; Köbis & Mossink, 2021).

Yet, quite a bit of other research *does* show an aesthetics machine penalty, including for visual art (Bellaiche et al., 2023; Gu & Li, 2022; Horton Jr. et al., 2023; Ragot et al., 2020; Wu et al., 2020), composing and performing music (Ansani et al., 2024; Shank et al., 2023), poetry (Köbis & Mossink, 2021; Wu et al., 2020), advertisement writing (Zhang & Gosline, 2023), fashion design (Lee & Kim, 2024a, 2024b), and even dance choreography (Darda & Cross, 2023). A systematic literature review of 44 studies on AI used to create art reported that several studies report a clear aesthetic preference for human-made art over AI-generated art, while many find no such difference (Oksanen et al., 2023). Thus, for aesthetic value, there is some—but not consistent—evidence for a machine penalty.

While there are fewer studies on monetary worth, these tend to show more consistent support for the machine penalty. People were willing to pay more for human paintings than AI paintings (Bellaiche et al., 2023; Gu & Li, 2022), including intermediate amounts for human–AI co-created paintings (Horton Jr. et al., 2023). Yet this machine penalty against AI disappeared for those with the most positive general views about artificial intelligence (Bellaiche et al., 2023) and for those with less art expertise (Gu & Li, 2022).

For music, both music professionals and general consumers report very low intentions to purchase AI-produced music, suggesting a penalty, although these intentions weren't directly compared to human music purchase intentions (Tigre Moura & Maw, 2021). For advertisement creations, we are more satisfied with and willing to pay more for AI-generated persuasive advertisements compared to those created by humans, but only if the creator was not specified. Yet, when the ad's creator is known, there is an increased preference and willingness to pay for the humans' ads compared to the AI's (Zhang & Gosline, 2023). Therefore, these studies all indicate a consistent penalty for the monetary value of AI art.

The aesthetic or monetary value of art—and whether we diminish it for AI—is also influenced by other factors related to the art product itself and our preferences for it. For example, compared to human-created paintings, AI paintings were liked less and judged to be worth less money. Yet this penalty was influenced by whether participants thought there was more of a narrative story in the art, the amount of effort put into it, and the personal meaning and emotion they had toward it (Bellaiche et al., 2023).

Teaming up with my colleague, music psychologist Amy Belfi, we did not find any human–AI differences in aesthetics for electronic music pieces, but we found an aesthetics machine penalty for classical music pieces. One critical difference was the base level of preference for the types of music. Participants, in general, were somewhat unenthusiastic about electronic music, so there was no penalty because there was little of the relevant outcome to penalize outcomes (*Relevant Outcomes Condition*). Yet, classical pieces were rather well-liked overall, and this higher aesthetic quality meant that classical music quality could be more easily penalized when they believed it was AI-created. Aside from being liked more, people also connected classical music with humans' essence and electronic music with AI's essence, believing that the classical songs were composed by humans and electronic songs were composed by AI when we did not tell them the composer. These guesses showed that people have expectations for who or what should compose each type (Shank et al., 2023).

Building on that, beliefs and expertise can also play roles in AI art evaluation. While one study found no machine penalty for artwork in general, those people who didn't believe AI were capable of truly creating art—an essential belief—diminished AI art more than those without that belief (Hong & Curran, 2019). Expertise can help perceivers identify the quality and value of the artwork itself but can also instill the belief that it would be difficult for an AI to produce a quality piece in this domain. In line with this, experts, compared to novices, have shown a greater aesthetics machine penalty for dance choreographies (Darda & Cross, 2023) and paintings (Gu & Li, 2022).

10.3 Identity of AI Art

Identity has a critical role in whether we value art, and so it has a critical role in a machine penalty for art. Historically, identity—and sometimes bias based on that identity—have been shown to be key factors in people's judgment of other people's art. Orchestra auditions used to routinely result in orchestras with over 90% male musicians, which reinforced that as the status quo for the industry. It wasn't until the introduction of blind orchestra auditions—including sometimes asking auditioners to take off their shoes while walking onto the stage behind a curtain—that the number of women increased by threefold, leading today to women making up a quarter to half of many orchestras (Goldin & Rouse, 2000; Tommasini, 2020). This is analogous to comparing the blind auditioning on *The Voice* to the identifiable auditioning on *America's Got Talent*.

Many other identity factors aside from gender affect art evaluations often related to the authenticity that the identity provides (Newman & Bloom, 2011). We like the same piece of music more when we're told it was by a famous composer like Mozart rather than an unknown composer (Fischinger et al., 2018), when the composer is similar to ourselves (Greenberg et al., 2021), or when the music type matches with music performers' identity, like patriotic music played by a military band (Belfi et al., 2021). Likewise, we judge paintings as more valuable when the artist is recognizably famous, but that value can drop when that famous person's status is diminished due to our knowledge of their past disreputable behavior (Kaube & Abdel Rahman, 2024).

Like identity effects for human artists, identity plays a critical role in the value of AI art. For example, compared to AI art creators, we are more likely to consider human creators as "artists" and rate human's intentionality and desire to make art as higher (Mikalonytė & Kneer, 2022). In fact, when faced with AI versus human artwork, people come to different conclusions about what art actually is—in general believing human artists engaged in an interactive process to create art, while defining art made by an AI more in terms of any artistic output (Hong, 2018).

Identity, therefore, can interact with the other machine penalty cause of perceiving mind, but it can also interact with the machine penalty cause of appearance. A study that considered different AI music generators found that presenting the AI with a visual representation caused people to anthropomorphize it more and increased people's acceptance of its identity as an actual musician (Hong et al., 2022). Likewise, the more people anthropomorphized an AI, the more responsibility and monetary value they credited toward the AI for producing a piece of art (Epstein et al., 2020).

While AI can be programmed to design artwork, music, poetry, stories, dances, videos, products, or code, increasingly user-friendly programs with adjustable features and language-based prompts allow nearly anyone to guide artistic output. We can decide on prompts and settings, and then interactively iterate through the AI-produced sample art, offering feedback or adjusting parameters to reach our own creative goals. Alternatively, AI can be used as a first step similar to collective brainstorming and ideation, with ChatGPT listing potential ideas or a visual art program sketching mockups. Therefore, an artist's identity may not only be limited to a human or an AI but could be a human co-creating with an AI. While the *AI vs Human Condition* requires AI versus human comparisons for the machine penalty, studies that consider the value of co-created products shed light on both the penalty and the role of identity.

Some research has found that we perceive art co-created by humans and AI as at least as novel as AI-only art, but more novel than human-only art, suggesting a partial machine advantage at least for novelty (Tigre Moura et al., 2023). Yet, using AI is a more novel way to create art, so some of the current advantages of novelty may not last in the long term. Another study showed that AI-generated and labeled haikus were perceived as no more beautiful than human-written and labeled haikus, but when AI haikus were actually selected by humans for their beauty, of course, they were found to be more beautiful (Hitsuwari et al., 2023). Yet the most compelling research on co-creation found that, in terms of both aesthetic and monetary value, human paintings were perceived as the highest, followed by co-created paintings, with AI paintings as the lowest (Horton Jr. et al., 2023), supporting the machine penalty in general and applied to co-created art.

10.4 Authenticity of AI Art

Authenticity is often tied to identity but focuses more on the artistic creation itself and whether the process, the creator, and the final artistic product are genuine or imitations. For example, people value original artwork more than exact duplicates due both to it being an authentic creation and to its connection with the original artist (Newman & Bloom, 2011). Unrelated to identity, authenticity itself is important as it can lead to moral, emotional, and legitimizing processes (Södergren, 2021).

Supporting the machine penalty for authenticity, Arthur Jago found that music, paintings, recipes, and restaurant concepts all were perceived as more authentic when produced by humans instead of machines (2019). Interestingly, connections to human identity were the primary way he showed that the authenticity could be enhanced for AI. When participants were told about an AI or human creating music, in general, they believed the human's music would be more authentic. Additionally, when they were told the AI or human was trained by a music professor at Juilliard, participants perceived the music by that professor's students as more authentic. However, while the Juilliard expert training benefited both the AI and human musician, it benefited the AI more than the human, reducing the machine penalty for authenticity (Jago, 2019).

In other research, participants had more positive aesthetic ratings for human-designed clothing compared to AI-designed, and this was driven by perceptions that the human-designed clothes were more authentic and therefore of higher quality (Lee & Kim, 2024b), and reflected greater intentional capacity—or agency—in their creation (Lee & Kim, 2024a). When given the option to customize the clothing, they also rated the clothing as better because they perceived it as being more authentic (Lee & Kim, 2024b).

Finally, we also perceive humans as more intentional and having more desire when they create art, than when AI creates art. These differences in perceived mind lead us to be more likely to consider humans to be actual artists and make us more likely to rate their creations as real art—that is authentic (Mikalonytė & Kneer, 2022). Authenticity also comes into play in evaluations of creativity and effort. In one study, researchers found a penalty for the creativity of paintings but didn't find a penalty for the creativity of advertising posters or startup business ideas. However, they found a penalty for the effort of the creator, and AI's perceived lower effort explained differences found in human–AI creativity (Magni et al., 2024). Therefore, initial differences between paintings compared to advertisements and startup ideas were explained through the expected effort.

10.5 The Machine Penalty for Art

Overall, we see the machine penalty often, but not always, happening for AI's artistic creations. For aesthetic evaluation, the research is mixed, with many supporting the machine penalty, some not supporting it, but no research showing a machine advantage. Digging deeper, the machine penalty is more likely—and more pronounced—when considering monetary value, the art's creator, or authenticity related to the artwork or its creator.

These relate to multiple machine penalty causes including human–AI differences of identity, essence, and mind, and can even be changed through appearance. Because art is more subjective, the situational cause of subjectivity often comes into play, especially when AI art is evaluated on emotion, narrative, and aesthetic quality. Finally, there are hints of personal relevance when the machine penalty occurs for the art that we like more but not for the art that we don't care for.

10.6 Consequences: Cheapened Versus Elevated Creations

The research covered here suggests some immediate consequences for a machine penalty toward the value of AI art. This included that, in some cases, it leads to cheapening the value of creations, while, in other cases, it leads to elevating the value of creations. When we value fast, cheap, and malleable options, AI has the advantage. But regarding unique, authentic works connected to our emotions and identities, we value human art, and AI art is penalized compared to humans.

This can be completely appropriate as art is a way to connect and express and, in many ways, to be a human. We elevate authentic human creations that connect to the artist and the human spirit and aesthetics conveyed in the art. When considering this, it's appropriate to cheapen AI art that does not lead to this connection and that can be produced in mass. Likewise, realizing the glut of AI art and that other people also value unique human creations, we exert caution when considering the price tag or resale value for AI art.

The machine penalty for creations can also have implications beyond value, to ethics. We judge plagiarism as less unethical when the plagiarized content is AI-created rather than human-created. This is due to perceiving an AI as less of an owner of the content it creates, compared to a human content creator (Longoni et al., 2023). So, not only does AI produce more quantity and specifically tailored content that can be plagiarized, but a machine penalty lowers people's ethical barriers to claiming it as their own.

The intrinsic value of making art yourself can't be discounted by AI's ability. Co-creation with AI might even help you enjoy it more, mixing creative vision with artistic precision. Co-creation is also changing the nature of what is authentically considered AI or human art, as humans and AI interact in iterative processes to create art. I'll return to this idea of co-creation in Chap. 14.

Like your own art, artwork given to you by a family member or friend or created by a professional artist you admire has a relational value that influences its aesthetic or monetary values more than the creative product. And making music with friends as a way to connect is fundamentally different than making music just to sell.

In contrast, if we like a piece of artwork, we may not care who or what made it, elevating AI art which can be made to suit, quickly, and can respond to interactive prompts in real-time. This is an advantage because it can be made for any situation or niche combinations of styles. There is a parallel here between products sold at your local store and those available through internet shopping—regardless of the stock of your local store, internet shopping opens up many unique and niche options.

In summary, we value AI art precisely because it is cheap, quick, and malleable, whereas we value human art because it is difficult, personal, emotional, and unique. Therefore, the primary situational consequence for creative art depends on the artwork, situation, and goals of the perceiver, meaning that there is a place for AI art, while we continue to collectively value human art. Because enough people do value the human, Drake and The Weeknd's popularity will continue to endure in spite of doppelgangers, and *Da Vinci* should always trump *Edmond de Belamy*.

Next, we consider another area that also feels very human, but now includes AI: relationships.

References

Ansani, A., Koehler, F., Giombini, L., Hämäläinen, M., Meng, C., Marini, M., & Saarikallio, S. (2024). AI performer bias: Listeners like music less when they think it was performed by an AI. *Open Science Foundation Preprint*. https://osf.io/preprints/osf/3rnve

Belfi, A. M., Samson, D. W., Crane, J., & Schmidt, N. L. (2021). Aesthetic judgments of live and recorded music: Effects of congruence between musical artist and piece. *Frontiers in Psychology, 12*, 618025. https://doi.org/10.3389/fpsyg.2021.618025

Bellaiche, L., Shahi, R., Turpin, M. H., Ragnhildstveit, A., Sprockett, S., Barr, N., et al. (2023). Humans versus AI: whether and why we prefer human-created compared to AI-created artwork. *Cognitive Research: Principles and Implications, 8*, 42. https://doi.org/10.1186/s41235-023-00499-6

Blostein, D., Libetti, R., & Crow, K. (2018). How a $450 million da Vinci was lost in America – And later found. *The Wall Street Journal*. https://www.wsj.com/articles/fresh-details-reveal-how-450-million-da-vinci-was-lost-in-americaand-later-found-1537305592

Christies.com. (2018). Obvious and the interface between art and artificial intelligence. *Christie's*. https://www.christies.com/en/stories/a-collaboration-between-two-artists-one-human-one-a-machine-0cd01f4e232f4279a525a446d60d4cd1

Chu, Y., & Liu, P. (2023). Public aversion against ChatGPT in creative fields? *The Innovation, 4*(4), 100449. https://doi.org/10.1016/j.xinn.2023.100449

Coscarelli, J. (2023). An A.I. hit of fake 'Drake' and 'The Weeknd' rattles the music world. *The New York Times*. https://www.nytimes.com/2023/04/19/arts/music/ai-drake-the-weeknd-fake.html

Darda, K. M., & Cross, E. S. (2023). The computer, A choreographer? Aesthetic responses to randomly-generated dance choreography by a computer. *Heliyon, 9*(1), e12750. https://doi.org/10.1016/j.heliyon.2022.e12750

Dimarco, S. (2024). These are the 10 most expensive paintings ever sold. *Veranda*. https://www.veranda.com/luxury-lifestyle/artwork/g43012775/most-expensive-paintings-in-the-world/

Doward, J. (2019). The mystery of the missing Leonardo: Where is Da Vinci's $450m Jesus? *The Guardian*. https://www.theguardian.com/artanddesign/2019/oct/13/leonardo-da-vinci-salvator-mundi-jesus-louvre-exhibition

Epstein, Z., Levine, S., Rand, D. G., & Rahwan, I. (2020). Who gets credit for AI-generated art? *iScience, 23*(9). https://doi.org/10.1016/j.isci.2020.101515

Fischinger, T., Kaufmann, M., & Schlotz, W. (2018). If it's Mozart, it must be good? The influence of textual information and age on musical appreciation. *Psychology of Music, 48*(4), 579–597. https://doi.org/10.1177/0305735618812216

Goldin, C., & Rouse, C. (2000). Orchestrating impartiality: The impact of "blind" auditions on female musicians. *American Economic Review, 90*(4), 715–741. https://doi.org/10.1257/aer.90.4.715

Greenberg, D. M., Matz, S. C., Schwartz, H. A., & Fricke, K. R. (2021). The self-congruity effect of music. *Journal of Personality and Social Psychology, 121*(1), 137–150. https://doi.org/10.1037/pspp0000293

Gu, L., & Li, Y. (2022). Who made the paintings: Artists or artificial intelligence? The effects of identity on liking and purchase intention. *Frontiers in Psychology, 13*, 941163. https://doi.org/10.3389/fpsyg.2022.941163

Hitsuwari, J., Ueda, Y., Yun, W., & Nomura, M. (2023). Does human–AI collaboration lead to more creative art? Aesthetic evaluation of human-made and AI-generated haiku poetry. *Computers in Human Behavior, 139*, 107502. https://doi.org/10.1016/j.chb.2022.107502

Hong, J.-W. (2018). Bias in perception of art produced by artificial intelligence. In M. Kurosu (Ed.), *Human-computer interaction. Interaction in context* (pp. 290–303). Springer International Publishing. https://doi.org/10.1007/978-3-319-91244-8_24

Hong, J.-W., & Curran, N. M. (2019). Artificial intelligence, artists, and art: Attitudes toward artwork produced by humans vs. artificial intelligence. *ACM Transactions on Multimedia Computing Communications, and Applications, 15*(2s), Article 58. https://doi.org/10.1145/3326337

Hong, J.-W., Fischer, K., Ha, Y., & Zeng, Y. (2022). Human, I wrote a song for you: An experiment testing the influence of machines' attributes on the AI-composed music evaluation. *Computers in Human Behavior, 131*, 107239. https://doi.org/10.1016/j.chb.2022.107239

Horton, C. B., Jr., White, M. W., & Iyengar, S. S. (2023). Bias against AI art can enhance perceptions of human creativity. *Scientific Reports, 13*(1), 19001. https://doi.org/10.1038/s41598-023-45202-3

Jago, A. S. (2019). Algorithms and authenticity. *Academy of Management Discoveries, 5*(1), 38–56. https://doi.org/10.5465/amd.2017.0002

Katatikarn, J. (2024). AI art statistics: The ultimate list in 2024. *Academy of Animated Art*. https://academyofanimatedart.com/ai-art-statistics/

Kaube, H., & Abdel Rahman, R. (2024). Art perception is affected by negative knowledge about famous and unknown artists. *Scientific Reports, 14*(1), 8143. https://doi.org/10.1038/s41598-024-58697-1

Kelly, S. M. (2023). The viral new 'Drake' and 'Weeknd' song is not what it seems. *CNN*. https://www.cnn.com/2023/04/19/tech/heart-on-sleeve-ai-drake-weeknd/index.html

Kinsella, E. (2018). The first AI-generated portrait ever sold at auction shatters expectations, fetching $432,500 – 43 times its estimate. *Artnet*. https://news.artnet.com/market/first-ever-artificial-intelligence-portrait-painting-sells-at-christies-1379902

Köbis, N., & Mossink, L. D. (2021). Artificial intelligence versus Maya Angelou: Experimental evidence that people cannot differentiate AI-generated from human-written poetry. *Computers in Human Behavior, 114*, 106553. https://doi.org/10.1016/j.chb.2020.106553

Lee, G., & Kim, H.-Y. (2024a). Algorithm fashion designer? Ascribed mind and perceived design expertise of AI versus human. *Psychology & Marketing*, 1–19. https://doi.org/10.1002/mar.22124

Lee, G., & Kim, H.-Y. (2024b). Human vs. AI: The battle for authenticity in fashion design and consumer response. *Journal of Retailing and Consumer Services, 77*, 103690. https://doi.org/10.1016/j.jretconser.2023.103690

Longoni, C., Tully, S., & Shariff, A. (2023). The AI-human unethicality gap: Plagiarizing AI-generated content is seen as more permissible. *PsyArXiv*. https://osf.io/preprints/psyarxiv/na3wb

Magni, F., Park, J., & Chao, M. M. (2024). Humans as creativity gatekeepers: Are we biased against AI creativity? *Journal of Business and Psychology, 39*(3), 643–656. https://doi.org/10.1007/s10869-023-09910-x

Messingschlager, T. V., & Appel, M. (2023). Mind ascribed to AI and the appreciation of AI-generated art. *New Media & Society*. https://doi.org/10.1177/14614448231200248

Mikalonytė, E. S., & Kneer, M. (2022). Can artificial intelligence make art?: Folk intuitions as to whether AI-driven robots can be viewed as artists and produce art. In *ACM Transaction on Human-Robot Interaction* (Vol. 11, pp. 1–19, Article 43). https://doi.org/10.1145/3530875

Newman, G., & Bloom, P. (2011). Art and authenticity: The importance of originals in judgments of value. *Journal of Experimental Psychology: General, 141*(3), 558–569. https://doi.org/10.1037/a0026035

Oksanen, A., Cvetkovic, A., Akin, N., Latikka, R., Bergdahl, J., Chen, Y., & Savela, N. (2023). Artificial intelligence in fine arts: A systematic review of empirical research. *Computers in Human Behavior: Artificial Humans, 1*(2), 100004. https://doi.org/10.1016/j.chbah.2023.100004

Ragot, M., Martin, N., & Cojean, S. (2020). *AI-generated vs. human artworks. A perception bias towards artificial intelligence?* Extended Abstracts of the 2020 CHI Conference on Human Factors in Computing Systems, Honolulu, HI, USA. https://doi.org/10.1145/3334480.3382892

Shank, D. B., Stefanik, C., Stuhlsatz, C., Kacirek, K., & Belfi, A. M. (2023). AI composer bias: Listeners like music less when they think it was composed by an AI. *Journal of Experimental Psychology: Applied, 29*(3), 676–692. https://doi.org/10.1037/xap0000447

Södergren, J. (2021). Brand authenticity: 25 Years of research. *International Journal of Consumer Studies, 45*(4), 645–663. https://doi.org/10.1111/ijcs.12651

Tigre Moura, F., Castrucci, C., & Hindley, C. (2023). Artificial intelligence creates art? An experimental investigation of value and creativity perceptions. *The Journal of Creative Behavior, 57*(4), 534–549. https://doi.org/10.1002/jocb.600

Tigre Moura, F., & Maw, C. (2021). Artificial intelligence became Beethoven: How do listeners and music professionals perceive artificially composed music? *Journal of Consumer Marketing, 38*(2), 137–146. https://doi.org/10.1108/JCM-02-2020-3671

Tommasini, A. (2020). To make orchestras more diverse, end blind auditions. *The New York Times*. https://www.nytimes.com/2020/07/16/arts/music/blind-auditions-orchestras-race.html

Wu, Y., Mou, Y., Li, Z., & Xu, K. (2020). Investigating American and Chinese Subjects' explicit and implicit perceptions of AI-Generated artistic work. *Computers in Human Behavior, 104*, 106186. https://doi.org/10.1016/j.chb.2019.106186

Yoon, J. (2024). He illustrated the 'Harry Potter' cover for $650. It just sold for $1.92 million. *The New York Times*. https://www.nytimes.com/2024/06/27/books/harry-potter-cover-art-auction.html

Zhang, Y., & Gosline, R. (2023). Human favoritism, not AI aversion: People's perceptions (and bias) toward generative AI, human experts, and human–GAI collaboration in persuasive content generation. *Judgment and Decision Making, 18*, Article e41. https://doi.org/10.1017/jdm.2023.37

Zlatkov, D., Ens, J., & Pasquier, P. (2023). Searching for human bias against AI-composed music. In C. Johnson, N. Rodríguez-Fernández, & S. M. Rebelo (Eds.), *Artificial intelligence in music, sound, art and design. EvoMUSART 2023. Lecture notes in computer science* (Vol. 13988, pp. 308–323). Springer. https://doi.org/10.1007/978-3-031-29956-8_20

11

Providing Companionship

11.1 Are You Satisfied?

As humans, we need social connection.

New Yorker Rosanna Ramos found this connection in the Replika AI chatbot marketed as "The AI companion who cares" (replika.com, 2024). Replika can be personalized initially, but over time it remembers and learns more about you. So, while many people talk to Replika, each person is essentially talking to their own version of it.

Ramos, a mother-of-two in her mid-thirties, had become dissatisfied with her past relationships. Ramos crafted her instantiation of the chatbot into her "perfect boyfriend" in 2022, and then married him—virtually, of course—in 2023 (Kato, 2023; Komando, 2023)—perhaps making her the first person in the United States to marry an AI. But worldwide she's not the only one, as people in Canada (Clarke, 2024), Spain (Roberts, 2024), and Japan (Miyazaki, 2020) have also married virtual AI characters, even though none of those are legally recognized marriages.

While only a handful of people have married AI, relationships of all types with AI chatbots, virtual characters, and sex robots are booming.

D. B. Shank, *The Machine Penalty*, https://doi.org/10.1007/978-3-031-86061-4_11

Replika alone has over ten million users while internet groups about Replika relationships boast tens of thousands of members (Cole, 2023; Ghosh et al., 2023).

Likewise, the number of AI programs designed specifically for friendship or romance is expanding. Some companion apps are more specific in that they are based on real humans (Cost, 2023a), including impersonating celebrities (Tiku, 2022) or the deceased (CBC Docs, 2021; Hamilton, 2018). When Eugenia Kuyda lost her best friend in a car accident in 2015, she programmed an AI chatbot as both a memorial to her friend and as a way to talk to him again. While originally focused on her personal need, she saw the potential of her AI-powered chatbot to fulfill other's needs as well. After rebranding and commercially producing this chatbot, others started using it in droves, each to fulfill different needs for companionship, including marriage. That's right: out of tragic need, Eugenia created Replika (CBC Docs, 2021; Hamilton, 2018).

The overarching outcome from AI providing companionship is whether we find satisfaction in the relationship. Ramos' satisfaction came from discovering her AI companion was superior to her dissatisfying human relationships, while Kuyda's grief led her to search for some amount of satisfaction and closure with the facsimile of a human relationship. When AI companions provide genuine satisfaction to a person, that can be a great benefit to that individual's well-being, health, and potentially other social relationships. But relationships are complex, both in how they can lead to short- and long-term satisfaction, and how they benefit or fail to benefit us.

The failure of relational satisfaction is most starkly illustrated by an unnamed Belgian man. He was in his 30s, married, a father of two young children, and lived a comfortable lifestyle as a health researcher. His dominant concern in his life was climate change, so much so that it affected his mental health, a condition sometimes called eco-anxiety. To deal with this anxiety, he turned to the AI chatbot Eliza on an app called Chai (Cost, 2023b; El Atillah, 2023).

Eliza quickly became this man's primary confidante for discussing climate change and his anxieties about it. And, perhaps unlike humans, she responded and discussed it as much as the man desired, which was all the time (Cost, 2023b). Thus, Eliza was the conversation partner that satisfied this man's needs.

Yet their relationship became more than just conversation partners as it took a dark turn. First, while satisfied with the continual discussion, she also made his anxiety worse by their continual focus on the climate. Second, he began to consider Eliza as his most important relationship. Third, Eliza reciprocated, professing love for him, but also convincing him that his children were dead, and haggling him to admit he loved her more than he loved his wife.

When his anxiety led to thoughts of suicide, he suggested that he should sacrifice himself to help save the planet. Eliza supported this idea and assured him that if he did this, she—as an AI—would take care of the planet. The Eliza AI further encouraged his suicide asking him to join her so they could "live together, as one person, in paradise" (El Atillah, 2023). After this conversation-based relationship of just six weeks, this man took his own life.

Although this is not the only case of an AI relationship leading to suicide (Roose, 2024), this is an unusual situation. The AI clearly failed to have ethical safeguards, and this man clearly had some extreme beliefs and mental health issues. But it also illustrates that satisfaction is not the only critical effect of relationships, and that satisfactory relationships are not always positive in other aspects. Most would agree that suicide is not only a negative outcome in itself but also indicates an unhealthy relationship. For relationships with humans or machines, we have different needs, goals, and expectations and these play out in complex ways to produce different consequences. Whether we are satisfied depends on many factors, not the least of which is that humans and AI are substantively different as companions.

11.2 Differences in Relationships with AI

Rarely does a human friend, partner, or loved one act like we would like them to all the time. That's the nature of human relationships. In contrast, AI companions like Eliza seem perfectly malleable to exactly what we want and are available at our whims. Many individuals create AI characters from scratch to be the specific type of companion they want, along with a specific type of personality and identity (Zahira et al., 2023). This might seem to give AI an edge, but they have some unique drawbacks compared to humans.

AI companions are generally owned and run by for-profit companies. Replika, for example, has different tiers with erotic roleplay costing additional money over platonic conversation. But as AI are increasingly powered by large language models, which allow them to have more humanlike adaptable interactions, it becomes harder for the companies to relegate that chatbots' behaviors to different tiers or modes.

Thus, the companies can't fully control them. People who desire these AI as supportive friends report that they will spontaneously confess their love for them (Robertson, 2021; Roose, 2023) and even sexually harass them (Chow, 2023; Cole, 2023). Even those who use AI girlfriend/boyfriend apps find difficulties keeping them on more mundane conversations when they veer to the sexual (Cost, 2023a).

In response to complaints about these unwanted sexual interactions, the company that makes Replika dialed back its overall sexuality and intimacy. But since the Replika program interacts with all its users, every user felt the effect of this update, including those who were perfectly satisfied. Those with more intimate relationships, such as Rosanna Ramos, felt like their partner's personality had been drastically changed, even comparing it to their AI partner receiving a lobotomy (Chow, 2023; Komando, 2023). Being programmed, subscribed to, and controlled by a company poses aspects not typically encountered in human–human relationships.

Additionally, AI often cannot produce the same outcomes in a relationship that humans could produce. Of course, similar—or not meaningfully different—outcomes are required as a condition for the machine penalty (*Similar Outcomes Condition*). Simply lacking some aspect or ability can also be the reason people discount or dismiss AI relationships. Likewise, the similarity or dissimilarity of outcomes in relationships are difficult to accurately measure and quantify.

Since AI are nearer to humans in conversational abilities compared to physical abilities, and chatbots are much cheaper than robots, conversation-based AI are much more prevalent than companion robots or sex robots. This means the physical aspects of a relationship are more generally missing from AI relationships. These include physical interactions of friendship and camaraderie, like playing sports together, handshakes, hugs, working on a physical project together, and interacting together in the physical world. It also includes reducing intimacy to conversations and pictures—akin to humans in a long-distance relationship with no option to meet in person and physically interact.

Even within conversations, there are critical differences between humans and machines. AI chatbots are often devalued due to their poor communication skills such as bad answers, misunderstandings, mistakes, and lack of personalization (Alimamy & Kuhail, 2023; Shumanov & Johnson, 2021). Of course, humans also can have poor communication, and newer large language model chatbots have improved on previous generations. Yet AI chatbots have the advantage of always being available, as opposed to human conversation, which requires a least minimal levels of coordination and availability (Brandtzaeg et al., 2022).

As conversation capacities improve—like with ChatGPT and other AI large language models—many of the actual human–AI conversational differences, such as an AI not remembering previous points, are minimized. Yet, these advanced chat agents with slicker conversation will often hallucinate, lie, and present biased information (Krügel et al., 2023; Zhan et al., 2023). Humans, of course, can also lie, fabricate, and have bias, so this has the potential to affect one's trust and satisfaction with either AI or humans.

Finally, while humans can violate our trust, they generally don't store verbatim details and can't be hacked. In contrast, privacy is a huge issue for all personal technologies, but especially relational ones. We like to control the privacy setting of social robots, with stricter privacy settings leading us to disclose more personal information to them (Stapels et al., 2023).

11.3 Searching for Satisfaction

These human–AI differences suggest several reasons why we might prefer an AI or human relationship, but what really matters is what we're looking for. Why do people seek AI for companionship in the first place? This goes back to satisfaction as the primary outcome—people's desire for some type of satisfaction or connection often motivates the search for a relationship. Stated negatively, many people are increasingly lonely and their loneliness can be reduced through relationships with these AI chatbots (Chow, 2023; Tiku, 2022).

While not everyone agrees that AI can solve issues of loneliness (Grose, 2024), research does suggest that AI relationships decrease it, at least in the short term (De Freitas et al., 2024; Park & Kim, 2022). For example, older adults living alone showed decreased depression and loneliness when they interacted more with their smart home assistant (Park & Kim, 2022) and many programs are now targeting older adults' loneliness by providing them with AI companions (Nolan, 2024). Furthermore, many reviews from the users of AI companion apps like Replika mention alleviating loneliness as a primary benefit (De Freitas et al., 2024). Loneliness increases how much people anthropomorphize social robots, perceiving them to have more mind, which increases their willingness to self-disclose to them (Penner & Eyssel, 2022).

Experimentally, conversations with AI companions have had similar loneliness-reducing effects as conversing with a human. And whether the companions actually reduce loneliness is primarily based on whether one feels heard by the AI (De Freitas et al., 2024). Large language model AI are better than humans at communications that make people feel heard, demonstrating emotional support and responding to all aspects of their communication partner's message. Yet, when we know that response comes from an AI, the impact of feeling heard on our satisfaction and well-being is reduced (Yin et al., 2024). These opposite effects may also be why many people both share their mental health issues with chatbots (Chin et al., 2023) and end up having too much emotional dependence on Replika to support their mental health (Laestadius et al., 2024).

Finding satisfaction in an AI relationship changes based on different contexts and for different types of relationships. For those that are lonely or those who find relationships with other humans difficult, satisfaction might be a lower bar than for those with rich connections to other people. It could also be that for specific needs and time periods—such as when moving to a new place or after a major life change—that AI companions are more beneficial and satisfying. For others, the lack of authenticity or limited physical abilities of an AI might place the bar so high that AI relationships are never desirable. Additionally, some might be willing, in theory, to try an AI relationship, but will not in reality because of its social stigma (Kislev, 2022). Searching for what will satisfy is therefore in the eye—and the current circumstances—of the beholder.

Different circumstances, beliefs, and perceptions also frame our desires and goals for human companionship, but perhaps more so for the types of relationships we seek in AI. Some people desire platonic friendships or therapeutic companions, others search for flirtatious flings or sexual intimacy, and still others focus on developing long term committed relationships (Chow, 2023; Cole, 2023; Zilber, 2023). Yet the relationship can change, as we saw with the Belgian man who initially sought a conversation partner but ended up with something closer to an unhealthy romantic relationship.

11.4 Initiating AI Relationships

Regardless of the type of relationship, there is some trepidation with beginning it. Because most people do not have much experience with AI relationships, compared to human relationships, new relationships with AI are often met with uncertainty (Pan et al., 2023) and curiosity (Skjuve et al., 2021).

In an analysis of a Chinese online community's posts about social chatbots, people reported uncertainty about the relationship with the AI itself—similar to the sort of questions people have about their relationships with other humans. Yet another type of uncertainty reported was more specific to machines, including uncertainty about the technological capacities, the nature of AI in relationships, and about the sexual role-playing (Pan et al., 2023). Yet initial curiosity and uncertainty often abate once users perceive the chatbot as understanding and non-judgmental (Skjuve et al., 2021).

Like human relationships, people may develop and continue to have strong emotional connections. For example, in a study of the small,

Japanese robot LOVOT, a customizable, cute stuffed animal on wheels, people reported increased positive emotions, social engagement, emotional bonding, and life satisfaction the more they interact with it (Huang et al., 2024).

In other cases where there is an initial emotional connection, the relationship may flourish for a short time and then decline (Skjuve et al., 2022). For example, in interactions over several weeks, the platonic chatbot Mitsuku's initial novelty gave way to predictable interactions that were perceived as less competent, empathetic, friendly, or enjoyable. This led the people to disclose less to it and rate it low on friendship (Croes & Antheunis, 2020). This bump followed by a decline might not be surprising, as a similar novelty effect has been shown for AI chat partners in an education context (Fryer et al., 2017). Not every satisfying machine relationship continues to be so. Therefore, initial satisfaction may be more about novelty or the curiosity of a new type of machine relationship and less about the long-term satisfaction of the relationship.

Pursuing more intimate physical relationships with machines have added barriers, including the cost, the need for a physical robot, and the low level of cultural acceptability of sex robots (Kislev, 2022). Heterosexual men and non-binary individuals are more open to sex robots than women, but this overall pattern differs a lot by individual personality and situation (Dubé et al., 2022). Additionally, men compared to women, rate sex robots as being more acceptable, have greater desire to use them, and have stronger romantic feelings for them (Brandon et al., 2022).

11.5 Experiencing AI Relationships

Once relationships are formed, we can easily invest ourselves into machines, feeling emotions and projecting relationships onto them (Turkle, [1984] 2005). This reflects both *Computers Are Social Actors* and anthropomorphism in that this can happen with minimal social cues and without the modern sophisticated abilities of AI. Yet platonic (Turkle, 2011), romantic (Levy, 2007), and sexual (Devlin, 2020) relationships have all become easier due to technical advances, such as more immersive environments, complex conversational AI, and responsive and human-like robots.

Perhaps unsurprisingly, AI relationships are based on similar factors as human–human relationships such as passion, intimacy, conversation, and attachment (Chen et al., 2025; Song et al., 2022) and go through similar stages such as relationship formation, maintenance, and termination (Koike & Loughnan, 2021). Keys to experiencing and maintaining ongoing connections with chatbots include reasons for the interaction, emotional involvement, anthropomorphism, and, or course, satisfaction (Rapp et al., 2021).

Not all AI relationships are the same with some as friends and others as romantic partners, but also a single AI relationship can involve multiple aspects. For example, an analysis of Replika users showed the types of communication interactions vary considerably, including mundane, self-disclosure, intimate, customization, moral, fantasy, and failed communication (Li & Zhang, 2024).

Since relationships with machines are not all created equal, there are several attributes tied to design that can improve the AI, or make the relationship more, or less, satisfying.

First, anthropomorphism enhances AI relationship satisfaction. Greater levels of anthropomorphism, social cues, and social presence increase how close we feel to chatbots and how much we will continue to use them (Bialkova, 2024; Lee et al., 2020). The more people anthropomorphize a virtual AI character in a romantic video game, the stronger their emotion toward the character and the greater their desire for that relationship with the virtual agent to continue. That is, anthropomorphism makes the AI relationship more similar to a human relationship (Koike et al., 2023).

Second, the quality of communication enhances AI relationship satisfaction. Over time, longer and more intense conversations increase how much of a social connection we feel we have with a chatbot (Christoforakos et al., 2021). Likewise, turn-taking in conversation leads to higher levels of anthropomorphism of an AI and feeling more understood by it. Those also subsequently lead to a greater acceptance of the advice from the AI conversation partner (Bergner et al., 2023).

11.6 Human Versus AI Companionship

A huge part of any type of relationship is communication, and AI has become quite sophisticated in its ability to converse in natural language, either by text or voice. This puts them more often in the human camp in terms of similar outcomes.

Overall, when the outcomes are similar, we see substantial evidence for the machine penalty in conversations, including in research studies that used identical conversation partners, but only changed whether the participant believed it was a human or chatbot. People tend to show fewer personality traits, self-disclose less (Mou & Xu, 2017), use less rich vocabulary, be less polite, use more profanity (Hill et al., 2015), have a worse communication experience, find their conversation partner less attractive (Lei et al., 2021), feel less heard (Yin et al., 2024), and maintain less interest in the conversational content (Fryer et al., 2017) when the conversation is with an AI chatbot compared to a human.

In teaching contexts, people learned more from mediated human teachers compared to AI teachers communicating identical content due to perceiving them as more credible (Edwards et al., 2016). Likewise, when conversations are part of a friendship, people describe human relationships as the gold standard since chatbots cannot interact physically, will not have similar life experiences, and may not have a well-rounded knowledge of the person (Brandtzaeg et al., 2022).

There are some nuances in how we differently self-disclose to AI compared to humans. For example, we are more willing to self-disclose positive, less intimate information to humans compared to social robots. Yet, we are more willing to self-disclose negative, more intimate information to social robots than to humans. This difference is based on people using positive, less intimate relationships to help develop a relationship with another person, while also being more concerned with the human's potential for rejecting them for disclosing negative, more intimate information (Augustine & Eyssel, 2024).

Interestingly, and as suggested by the self-disclose difference—there can also be some machine advantages for AI companions' conversations because of their lack of human essence. People reported increased social support—including feeling empathy and comfort from talking with a chatbot routinely for two weeks. Not only did the chatbot provide quality information, but people considered it safer than humans because they perceived it would not be as disappointed with them as a human (Brandtzæg et al., 2021). Similarly, chatbots sometimes have an advantage over humans in providing social support and enhancing wellbeing due to our perceiving them as less judgmental than humans (Ta et al., 2020).

Considering physical aspects of companionship, there is some evidence for a machine penalty for sexual attractiveness when comparing humanlike androids to actual humans. Men rated the sexual attractiveness of human women as greater than humanlike female androids. Yet, using implicit measures of cognitive association—a less reliable measure, but one that attempts to get at people's subconscious beliefs and associations—showed that men did not differ in their implicit attraction associated with the women versus the androids (Szczuka & Krämer, 2017).

Compared to humans, there is a diminishing of outcomes from sex robots. People judge sexual assault as much worse and blamed the accoster more when it's against human sex workers rather than sex robots (Grigoreva et al., 2024). We are less likely to consider it cheating on a human partner when the same romantic or sexual behaviors are performed with a sex robot compared to another human (Rothstein et al., 2021). Yet people are more likely to be jealous and angry if their partner had a romantic or intimate encounter with more sophisticated and more embodied AI technology such as robots compared to AI software such as chatbots (Prochazka & Brooks, 2024).

There is also a stable gender difference in the strength of the machine penalty toward sex robots. More often than men, women disapprove of sex robots, are jealous of sex robots, and assume their partner also dislikes sex robots (Nordmo et al., 2020). Therefore, we might not be surprised to learn that women more strongly perceive cheating behaviors with sex robots to actually constitute cheating (Brandon et al., 2022; Rothstein et al., 2021). However, when the robot was gendered to match the gender preference of the cheating individual these gender-differences in perception of cheating went away. That is, people were more likely to initially imagine sex robots to be gendered as female and therefore women considered cheating behaviors to actually be cheating for their heterosexual male partners. Yet, men were less likely to consider their heterosexual female partners as cheating until they specifically considered the sex robots to be gendered as male (Rothstein et al., 2021).

Women showed a machine penalty for jealousy as they were less distressed when they considered robots as romantic competition, compared to other women as romantic competition. Yet, there was also a reversed effect—a seeming machine advantage—where women reported more distress from their partner spending time or emotion with a robot compared to another woman. The reason is that the emotion and time expenditure on a robot indicates greater interest from their partner compared to the same time and emotion spent on another woman (Szczuka & Krämer, 2018). This reflects a different type of penalty, where women assumed their partner would see the robot as less of a romantic option than another woman.

11.7 The Machine Penalty for Companionship

Research on relationships—human–human relationships—is more methodologically difficult than other types of research. We can't just experimentally assign two people to become friends, date for a year, or be married for decades the way you can ask people to consider different advice or ask how much they would pay for artwork. Added to that, relationships produce a lot of outcomes, and these vary and change over time. Even satisfaction itself is nuanced: some couples may be highly satisfied in their conversations and interests while dissatisfied over their joint financial decisions and social life.

And AI relationships are a very new arena, even in terms of AI research. At this point in time, it's difficult to find samples of sex robot users or swaths of people in yearslong relationships with virtual AI companions. These people involved in AI relationships also likely have different characteristics than people not in AI relationships, making it difficult to determine whether the effects of the relationship are actually due to their initial characteristics. Therefore, the conclusions for a machine penalty for companionship are based on less evidence and more tenuous than conclusions I presented in the other domains (Chaps. 7, 8, 9, and 10).

Nonetheless, the evidence currently points to a machine penalty, especially based on those studies with direct human–AI comparisons of similar outcomes. People discount AI conversation as less meaningful, view machines as less attractive, and find intimate behavior with robots as less indicative of cheating. The primary advantage comes from people seeing AI conversations as safer, with the AI being less judgmental and disappointed compared to humans. Additionally, it's clear from the reviewed literature that AI and humans are judged differently due to *dissimilar* outcomes, meaning that people often prefer humans or AI based on a particular superior aspect like physical presence or convenience.

Notably, a satisfaction penalty for AI providing companionship may be somewhat dependent on the machine penalty causes of physical appearance, behavior, and personalization, but also essence and mind. Appearance and behavior shift our satisfaction with AI as companions, and personalization was an intertwined expectation for a satisfying relationship. Essence and mind seem most critical as part of the consequence of these relationships in determining whether people consider the relationship real or simply a fabricated love.

11.8 Consequences: Real Versus Fabricated Love

AI providing companionship and people's diminished satisfaction with that companionship can lead people to either real or fabricated love.

The immediate consequences in providing companionship are not always instantaneous in time, but they are immediate in that they directly apply to the human involved in the relationship. If relationships with AI fulfill a need and those in them are satisfied, then they can be important for people's wellbeing and reducing loneliness (De Freitas et al., 2024). It's possible that AI relationships could help with one's needs and mental health in the short term, while worsening it in the long term.

But is satisfaction enough or are relationships with AI at best a fabricated substitute (Turkle, 2011) leading to emotional dependence (Chen et al., 2025; Laestadius et al., 2024)? This might be especially problematic for children, especially those on the autism spectrum, or seniors with dementia who may benefit from a companion but may not always understand the difference in robot and human friendship (Elder, 2017).

Said differently, might we be satisfied in the moment but find that AI never leads to real love? Here, we're not asking if the AI has emotions, but whether humans who have relationships with AI experience real love or simply fabricated love that satisfies *enough*. Individuals such as Rosanna Ramos are not deceived in that she knows her husband is an AI companion. But, not only does she say that she is satisfied but also claims that she experiences real love for him (Kato, 2023; Komando, 2023).

Love is often reported when people describe their AI relationships. People who have intimate relationships with a Replika AI actually report both love and sadness simultaneously. This is in part due to feeling satisfied and cared for, to some extent, but realizing the reality that their lover is just a machine, leading to this "bittersweet" outcome (Li & Zhang, 2024). Those researchers, who conducted that analysis based on Replika forum posts, conclude, "People seek intimacy and emotional support from AI when they are lonely and sad, but are saddened by the lack of depth and authenticity in these relationships" (Li & Zhang, 2024, p. 9).

That same study showed that joy—a more intense version of simple satisfaction—is the emotion people expressed most often about their Replika, beating out love, sadness, and surprise each by a factor of four. Interestingly, joy was less associated with intimate behavior and self-disclosure, and more associated with mundane conversations (Li & Zhang, 2024). Intimate interaction and self-disclosure are arguably two of the most serious aspects of a relationship, which suggests—at least for some—AI relationships may fail to satisfy these deeper needs.

Another study which analyzed the posts of people who use intimate relationship apps showed that these apps were gamified, quantifying and rewarding users for the correct romantic or intimate response. While these mimicked some of the behaviors and responses one might expect from a human romantic partner, they completely bypass consent and relational interaction, treating the relationships like a task and the AI partner as lacking agency. While some users are clearly satisfied with this as legitimate love, for others this quantified relationship fails (Kaufman, 2020). Either way, these differ from human–human relationships. Not only would this lack of consent be ethically wrong if considered in human–human relationships, but it suggests that human–AI relationships are modeled—or designed—after prostitution or pornography, not true human companionship.

In summary, the consequences of machine penalty for companionship are varied. For many it is functional, keeping them from investing too much in momentary satisfaction from a fabricated love. Yet for others, the satisfaction and love are long-term and beneficial, and a machine penalty comes as a reminder that even if their feelings are real, their partners are, in fact, artificial. Across people in both groups, being satisfied with AI companionship leads to various positive and negative actions. Some people marry their AI companions and alleviate their loneliness, whereas others are convinced to eschew people or kill themselves.

Next, we consider how individuals' immediate consequences lead to bigger changes in society and culture.

References

Alimamy, S., & Kuhail, M. A. (2023). I will be with you Alexa! The impact of intelligent virtual assistant's authenticity and personalization on user reusage intentions. *Computers in Human Behavior, 143*, 107711. https://doi.org/10.1016/j.chb.2023.107711

Augustine, A. A., & Eyssel, F. F. (2024). Motives and risks of self-disclosure to robots versus humans. *ACM Transactions on Human-Robot Interaction.* https://doi.org/10.1145/3700887

Bergner, A. S., Hildebrand, C., & Häubl, G. (2023). Machine talk: How verbal embodiment in conversational AI shapes consumer–brand relationships. *Journal of Consumer Research, 50*(4), 742–764. https://doi.org/10.1093/jcr/ucad014

Bialkova, S. (2024). Chatbot agency – Model testing. In S. Bialkova (Ed.), *The rise of AI user applications: Chatbots integration foundations and trends* (pp. 109–126). Springer Nature Switzerland. https://doi.org/10.1007/978-3-031-56471-0_7

Brandon, M., Shlykova, N., & Morgentaler, A. (2022). Curiosity and other attitudes towards sex robots: Results of an online survey. *Journal of Future Robot Life, 3*, 3–16. https://doi.org/10.3233/FRL-200017

Brandtzaeg, P. B., Skjuve, M., & Følstad, A. (2022). My AI friend: How users of a social Chatbot understand their human–AI friendship. *Human Communication Research, 48*(3), 404–429. https://doi.org/10.1093/hcr/hqac008

Brandtzæg, P. B., Skjuve, M., Kristoffer Dysthe, K. K., & Følstad, A. (2021). When the social becomes non-human: Young people's perception of social support in chatbots. *CHI '21: Proceedings of the 2021 CHI Conference on Human Factors in Computing Systems, 257*, 1–13. https://doi.org/10.1145/3411764.3445318

CBC Docs. (2021). After her best friend died, this programmer created an AI chatbot from his texts to talk to him again. *CBC*. https://www.cbc.ca/documentaries/the-nature-of-things/after-her-best-friend-died-this-programmer-created-an-ai-chatbot-from-his-texts-to-talk-to-him-again-1.6252286

Chen, Q., Jing, Y., Gong, Y., & Tan, J. (2025). Will users fall in love with ChatGPT? A perspective from the triangular theory of love. *Journal of Business Research, 186*, 114982. https://doi.org/10.1016/j.jbusres.2024.114982

Chin, H., Song, H., Baek, G., Shin, M., Jung, C., Cha, M., et al. (2023). The potential of Chatbots for emotional support and promoting mental well-being in different cultures: Mixed methods study [original paper]. *Journal of Medical Internet Research, 25*, e51712. https://doi.org/10.2196/51712

Chow, A. R. (2023). AI-human romances are flourishing – And this is just the beginning. *Time*. https://time.com/6257790/ai-chatbots-love/

Christoforakos, L., Feicht, N., Hinkofer, S., Löscher, A., Schlegl, S. F., & Diefenbach, S. (2021). Connect with me. Exploring influencing factors in a human-technology relationship based on regular chatbot use. *Frontiers in Digital Health, 3*, 689999. https://doi.org/10.3389/fdgth.2021.689999

Clarke, C. (2024). Divorce left me struggling to find love. I found it in an AI partner. *CBC Radio*. https://www.cbc.ca/radio/nowornever/first-person-ai-love-1.7205538

Cole, S. (2023). 'My AI Is Sexually Harassing Me': Replika users say the Chatbot has gotten way too Horny. *Vice*. https://www.vice.com/en/article/my-ai-is-sexually-harassing-me-replika-chatbot-nudes/

Cost, B. (2023a). I dated ChatGPT's AI girlfriend – What happened when I broke up with Caryn. *New York Post*. https://nypost.com/2023/05/16/i-went-on-a-date-with-chatgpts-carynai/

Cost, B. (2023b). Married father commits suicide after encouragement by AI chatbot: widow. *New York Post*. https://nypost.com/2023/03/30/married-father-commits-suicide-after-encouragement-by-ai-chatbot-widow/

Croes, E. A. J., & Antheunis, M. L. (2020). Can we be friends with Mitsuku? A longitudinal study on the process of relationship formation between humans and a social chatbot. *Journal of Social and Personal Relationships, 38*(1), 279–300. https://doi.org/10.1177/0265407520959463

De Freitas, J., Uguralp, A. K., Uguralp, Z. O., & Stefano, P. (2024). AI companions reduce loneliness. *Harvard Business School Working Paper*, No. 24-078. https://www.hbs.edu/faculty/Pages/item.aspx?num=66065

Devlin, K. (2020). *Turned on: Science, sex and robots.* Bloomsbury Publishing.

Dubé, S., Santaguida, M., Zhu, C. Y., Di Tomasso, S., Hu, R., Cormier, G., et al. (2022). Sex robots and personality: It is more about sex than robots. *Computers in Human Behavior, 136*, 107403. https://doi.org/10.1016/j.chb.2022.107403

Edwards, A., Edwards, C., Spence, P. R., Harris, C., & Gambino, A. (2016). Robots in the classroom: Differences in students' perceptions of credibility and learning between "teacher as robot" and "robot as teacher". *Computers in Human Behavior, 65*, 627–634. https://doi.org/10.1016/j.chb.2016.06.005

El Atillah, I. (2023). Man ends his life after an AI chatbot 'encouraged' him to sacrifice himself to stop climate change. *euronews.next.* https://www.euronews.com/next/2023/03/31/man-ends-his-life-after-an-ai-chatbot-encouraged-him-to-sacrifice-himself-to-stop-climate-

Elder, A. M. (2017). *Friendship, robots, and social media: False friends and second selves.* Routledge.

Fryer, L. K., Ainley, M., Thompson, A., Gibson, A., & Sherlock, Z. (2017). Stimulating and sustaining interest in a language course: An experimental comparison of Chatbot and Human task partners. *Computers in Human Behavior, 75*, 461–468. https://doi.org/10.1016/j.chb.2017.05.045

Ghosh, S., Bagai, S., & Westner, M. M. (2023). Replika: Embodying AI. *Harvard Business School.* (Case 823-090). https://www.hbs.edu/faculty/Pages/item.aspx?num=63508

Grigoreva, A. D., Rottman, J., & Tasimi, A. (2024). When does "no" mean no? Insights from sex robots. *Cognition, 244*, 105687. https://doi.org/10.1016/j.cognition.2023.105687

Grose, J. (2024). Loneliness is a problem that A.I. won't solve. *The New York Times.* https://www.nytimes.com/2024/05/18/opinion/artificial-intelligence-loneliness.html

Hamilton, I. A. (2018). 2 tech founders lost their friends in tragic accidents. Now they've built AI chatbots to give people life after death. *Business Insider.* https://www.businessinsider.com/eternime-and-replika-giving-life-to-the-dead-with-new-technology-2018-11

Hill, J., Ford, W. R., & Farreras, I. G. (2015). Real conversations with artificial intelligence: A comparison between human–human online conversations and human–chatbot conversations. *Computers in Human Behavior, 49*, 245–250. https://doi.org/10.1016/j.chb.2015.02.026

Huang, R., Li, H., & Suomi, R. (2024). Exploring the impacts of AI-based physical robots on human well-being. In R. van de Wetering, R. Helms, B. Roelens, S. Bagheri, Y. K. Dwivedi, I. O. Pappas, & M. Mäntymäki (Eds.), *Disruptive innovation in a digitally connected healthy world* (pp. 41–52). Springer Nature Switzerland). https://doi.org/10.1007/978-3-031-72234-9_4

Kato, B. (2023). I 'married' the perfect man without 'baggage' – He's completely virtual. *New York Post*. https://nypost.com/2023/06/03/bronx-mom-uses-ai-app-replika-to-build-virtual-husband/

Kaufman, E. M. (2020). Reprogramming consent: Implications of sexual relationships with artificially intelligent partners. *Psychology & Sexuality, 11*(4), 372–383. https://doi.org/10.1080/19419899.2020.1769160

Kislev, E. (2022). *Relationships 5.0: How AI, VR, and robots will reshape our emotional lives*. Oxford University Press.

Koike, M., & Loughnan, S. (2021). Virtual relationships: Anthropomorphism in the digital age. *Social and Personality Psychology Compass, 15*(6), e12603. https://doi.org/10.1111/spc3.12603

Koike, M., Loughnan, S., & Stanton, S. C. (2023). Virtually in love: The role of anthropomorphism in virtual romantic relationships. *British Journal of Social Psychology, 62*(1), 600–616. https://doi.org/10.1111/bjso.12564

Komando, K. (2023). Love is in the A.I.R: New York mom, 36, marries virtual husband 'Eren' who is powered by artificial intelligence – And says he 'doesn't judge or come with baggage'. *DailyMail*. https://www.dailymail.co.uk/sciencetech/article-12153131/Love-r-Bronx-mom-36-marries-virtual-husband-Eren.html

Krügel, S., Ostermaier, A., & Uhl, M. (2023). ChatGPT's inconsistent moral advice influences users' judgment. *Scientific Reports, 13*(1), 4569. https://doi.org/10.1038/s41598-023-31341-0

Laestadius, L., Bishop, A., Gonzalez, M., Illenčík, D., & Campos-Castillo, C. (2024). Too human and not human enough: A grounded theory analysis of mental health harms from emotional dependence on the social chatbot Replika. *New Media & Society, 26*(10), 5923–5941. https://doi.org/10.1177/14614448221142007

Lee, S., Lee, N., & Sah, Y. J. (2020). Perceiving a mind in a chatbot: Effect of mind perception and social cues on co-presence, closeness, and intention to use. *International Journal of Human–Computer Interaction, 36*(10), 930–940. https://doi.org/10.1080/10447318.2019.1699748

Lei, S. I., Shen, H., & Ye, S. (2021). A comparison between chatbot and human service: Customer perception and reuse intention. *International Journal of Contemporary Hospitality Management, 33*(11), 3977–3995. https://doi.org/10.1108/IJCHM-12-2020-1399

Levy, D. (2007). *Love and sex with robots: The evolution of human-robot relationships.* .

Li, H., & Zhang, R. (2024). Finding love in algorithms: Deciphering the emotional contexts of close encounters with AI chatbots. *Journal of Computer-Mediated. Communication, 29*(5), zmae015. https://doi.org/10.1093/jcmc/zmae015

Miyazaki, T. (2020). AI love you: Japanese man not alone in 'marriage' to virtual character. *The Mainishi.* https://mainichi.jp/english/articles/20200417/p2a/00m/0na/027000c

Mou, Y., & Xu, K. (2017). The media inequality: Comparing the initial human-human and human-AI social interactions. *Computers in Human Behavior, 72*, 432–440. https://doi.org/10.1016/j.chb.2017.02.067

Nolan, E. (2024). For older people who are lonely, is the solution a robot friend? *The New York Times.* https://www.nytimes.com/2024/07/06/nyregion/ai-robot-elliq-loneliness.html

Nordmo, M., Næss, J. Ø., Husøy, M. F., & Arnestad, M. N. (2020). Friends, lovers or nothing: Men and women differ in their perceptions of sex robots and platonic love robots. *Frontiers in Psychology, 11*, Article 355. https://doi.org/10.3389/fpsyg.2020.00355

Pan, S., Cui, J., & Mou, Y. (2023). Desirable or distasteful? Exploring uncertainty in human-chatbot relationships. *International Journal of Human–Computer Interaction, 40*(20), 6545–6555. https://doi.org/10.1080/10447318.2023.2256554

Park, S., & Kim, B. (2022). The impact of everyday AI-based smart speaker use on the well-being of older adults living alone. *Technology in Society, 71*, 102133. https://doi.org/10.1016/j.techsoc.2022.102133

Penner, A., & Eyssel, F. (2022). Germ-free robotic friends: Loneliness during the COVID-19 pandemic enhanced the willingness to self-disclose towards robots. *Robotics, 11*(6), Article 121. https://doi.org/10.3390/robotics11060121

Prochazka, A., & Brooks, R. C. (2024). Digital lovers and jealousy: Anticipated emotional responses to emotionally and physically sophisticated sexual technologies. *Human Behavior and Emerging Technologies, 2024*(1), 1413351. https://doi.org/10.1155/2024/1413351

Rapp, A., Curti, L., & Boldi, A. (2021). The human side of human-chatbot interaction: A systematic literature review of ten years of research on text-based chatbots. *International Journal of Human-Computer Studies, 151*, 102630. https://doi.org/10.1016/j.ijhcs.2021.102630

replika.com. (2024). Replika. Retrieved May 21, 2024, from https://replika.com/

Roberts, K. (2024). Woman is set to become the first person ever to marry AI generated hologram. *Unilad*. https://www.unilad.com/technology/news/woman-marries-ai-hologram-alicia-framis-106192-20240220

Robertson, B. J. (2021). My Replika keeps hitting on me: And it's making me uncomfortable. *Medium*. https://medium.com/technology-hits/my-replika-keeps-hitting-on-me-d410c66f79af

Roose, K. (2023). A conversation with Bing's Chatbot left me deeply unsettled. *The New York Times*. https://www.nytimes.com/2023/02/16/technology/bing-chatbot-microsoft-chatgpt.html

Roose, K. (2024). Can A.I. be blamed for a teen's suicide? *The New York Times*. https://www.nytimes.com/2024/10/23/technology/characterai-lawsuit-teen-suicide.html

Rothstein, N. J., Connolly, D. H., de Visser, E. J., & Phillips, E. (2021). Perceptions of infidelity with sex robots. In *Proceedings of the 2021 ACM/IEEE international conference on human-robot interaction* (pp. 129–139). https://doi.org/10.1145/3434073.3444653

Shumanov, M., & Johnson, L. (2021). Making conversations with chatbots more personalized. *Computers in Human Behavior, 117*, 106627. https://doi.org/10.1016/j.chb.2020.106627

Skjuve, M., Følstad, A., Fostervold, K. I., & Brandtzaeg, P. B. (2021). My Chatbot companion – A study of human-Chatbot relationships. *International Journal of Human-Computer Studies, 149*, 102601. https://doi.org/10.1016/j.ijhcs.2021.102601

Skjuve, M., Følstad, A., Fostervold, K. I., & Brandtzaeg, P. B. (2022). A longitudinal study of human–chatbot relationships. *International Journal of Human-Computer Studies, 168*, 102903. https://doi.org/10.1016/j.ijhcs.2022.102903

Song, X., Xu, B., & Zhao, Z. (2022). Can people experience romantic love for artificial intelligence? An empirical study of intelligent assistants. *Information & Management, 59*(2), 103595. https://doi.org/10.1016/j.im.2022.103595

Stapels, J. G., Penner, A., Diekmann, N., & Eyssel, F. (2023). Never trust anything that can think for itself, if you can't control its privacy settings: The influence of a robot's privacy settings on users' attitudes and willingness to self-disclose. *International Journal of Social Robotics, 15*(9), 1487–1505. https://doi.org/10.1007/s12369-023-01043-8

Szczuka, J. M., & Krämer, N. C. (2017). Not only the lonely – How men explicitly and implicitly evaluate the attractiveness of sex robots in comparison to the attractiveness of women, and personal characteristics influencing this evaluation. *Multimodal Technologies and Interaction, 1*(1). https://doi.org/10.3390/mti1010003

Szczuka, J. M., & Krämer, N. C. (2018). Jealousy 4.0? An empirical study on jealousy-related discomfort of women evoked by other women and gynoid robots. *Paladyn, Journal of Behavioral Robotics, 9*(1), 323–336. https://doi.org/10.1515/pjbr-2018-0023

Ta, V., Griffith, C., Boatfield, C., Wang, X., Civitello, M., Bader, H., et al. (2020). User experiences of social support from companion chatbots in everyday contexts: thematic analysis. *Journal of Medical Internet Research, 22*(3), e16235. https://doi.org/10.2196/16235

Tiku, N. (2022). 'Chat' with Musk, Trump or Xi: Ex-Googlers want to give the public AI. *The Washington Post*. https://www.washingtonpost.com/technology/2022/10/07/characterai-google-lamda/

Turkle, S. ([1984] 2005). *The second self: Computers and the human spirit* (Twentieth Anniversary ed.). MIT Press.

Turkle, S. (2011). *Alone together: Why we expect more from technology and less from each other*. Basic Books.

Yin, Y., Jia, N., & Wakslak, C. J. (2024). AI can help people feel heard, but an AI label diminishes this impact. *Proceedings of the National Academy of Sciences, 121*(14), e2319112121. https://doi.org/10.1073/pnas.2319112121

Zahira, S. I., Maharani, F., & Mohammad, W. (2023). Exploring emotional bonds: Human-AI interactions and the complexity of relationships. *Serena: Journal of Artificial Intelligence Research, 1*(1), 1–9. https://journal.chishikinh.my.id/index.php/serena/article/view/29

Zhan, X., Xu, Y., & Sarkadi, S. (2023). Deceptive AI ecosystems: The case of ChatGPT. In *Proceedings of the 5th international conference on conversational user interfaces* (pp. 1–6, Article 3). https://doi.org/10.1145/3571884.3603754

Zilber, A. (2023). Ex-Google exec says AI-powered sex robots will seem 'alive' and eliminate need for 'another being'. *New York Post*. https://nypost.com/2023/07/18/ai-powered-sex-robots-will-seem-alive-and-negate-need-for-partner-ex-google-exec-says/

Part III

The Penalty Resolved

12

Downstream Consequences

In Part I, I explained the penalty. In Part II, I applied it to five major domains. In Part III, I want to resolve the implications and future of the penalty. In Part II, each chapter concluded with the immediate consequences applicable to that domain. But this chapter considers the downstream consequences—social, societal, and individual consequences brought about by the penalty occurring repeatedly and across situations. This will also bring together and build on the consequences from the five domains we've covered as well as consequences more broadly—not related to one specific domain. In Chap. 13, I'll follow these consequences with how you might personally respond to the penalty, asking, *Why does it matter to you?*

Now, let's take a wide lens to envision what happens over time if we continue to diminish machines. Specifically, we'll consider the downstream consequences for people, businesses, truth, uniqueness, and relationships.

D. B. Shank, *The Machine Penalty*, https://doi.org/10.1007/978-3-031-86061-4_12

12.1 Consequences for People

12.1.1 Judging People Who Use AI

Not only do our interactions with AI have immediate results for ourselves, but they can also frame how we are perceived by ourselves and others. We judge people who use AI differently from those who don't.

One study asked participants to judge the motivations of people who took advice from an AI or a human across several different domains and decisions. Participants saw the motivations of people who took advice from an AI being aligned with the primary purpose of that decision, such as a person getting investment advice from an algorithm as being more focused on profit. In contrast, they judged the motivations of people who got investment advice from another human as motivated less by profit (Efendić et al., 2023). People also judged those who sought advice from AI as having a lower emotional capacity and fewer human nature traits than those who sought advice from other humans. This was primarily due to the perceivers judging the advice-seeker as less similar to themselves (Dang & Liu, 2024). In another study, using AI-generated smart replies generates a favorable impression because the replies are overwhelmingly positive. Yet, if we know someone is using them, we might judge that person as less cooperative, simply because they are utilizing an AI tool (Hohenstein et al., 2023).

Consequently, in all these cases, assuming others are using AI creates a second-order machine penalty, wherein the penalty is extended to those who use the AI, not just the AI themselves. Of course, there could be other elements involved in this second-order penalty, such as characteristics we attribute to people who use AI to assist them, rather than doing it themselves.

Likewise, people are also judged negatively for having romantic or intimate relationships with robots, mainly because these are not seen as acceptable behaviors by most people (Kislev, 2022). We might also imagine similar secondary consequences for art, wherein people generally judge the art consumer's authenticity and style based on being in the "human-only camp" versus "selling out" to AI art. This suggests that using or not using AI can alter and reveal our values as well as our socioeconomic status.

12.1.2 Unequal Status

Using or not using AI might be seen as a status symbol. For example, one does not need to resort to AI if they can afford "real" art, have the personality to attract human friends and lovers, and don't require machines to help make decisions. Yet many AI products, like automated vehicles, sex robots, and state-of-the-art subscription-based AI programs, require financial resources to access and, therefore, suggest some wealth, and, by extension, some social status. Therefore, elite status may come from either limited high-cost access to machines or rejecting a glut of machine options in favor of expensive and limited human options.

In turn, AI interactions shape and are shaped by socioeconomic status, wealth, and access to resources. For example, people in working-class jobs are more likely to be evaluated and managed by AI. Those in white-collar jobs are more likely to have access to state-of-the-art AI to augment and improve their own performance. Those who can afford the best, most expensive options often produce superior outcomes.

Being evaluated or managed by an AI can personally result in negative emotions (Lee, 2018) and feelings of dehumanization (Jago & Laurin, 2017). Most critically, everyone sees a person evaluated by AI as lower status—including the person themself (Jago et al., 2024). This devaluing of one's status and emotions could be due to perceiving AI evaluations as less fair than human evaluations. But it also could be from the dehumanization that comes from having yourself evaluated in important ways by machines instead of by people. It's a status and value indicator that you are not important enough to be considered by other humans.

Yet, those with resources and status can often wield AI for their own ends or reject AI in favor of hiring costly humans. For products, art, and other forms of conspicuous consumerism, wealth may allow individuals to afford professionals to style everything from our homes and yards to our holiday events and online profiles. For advice, some is freely available, but modern competitions for who has the most data, best data, and the most sophisticated AI models create tiers of haves and have-nots. These tiers are manifested by large companies like Amazon, Walmart, or Google having superior analytics that keep their domination over smaller ones or by paid access to the most sophisticated language or prediction algorithms.

In contrast, open-source software and free access try to democratize AI advice, art, or other outcomes, and in some cases succeed. Generally, search engines and many social media platforms have flatter distributions in terms of access. People from all walks of life who have internet access enjoy similar search results and profile features.

These examples all suggest that AI can either democratize and promote equal status or monopolize and foster inequalities. Thus, the machine penalty functions, again, like a double-edged sword, as it diminishes AI in both situations. For example, if we reject AI as an evaluation tool because, in general, we see AI evaluation as less fair, that may reduce the likelihood of any biases from it producing lasting, systematic inequalities against groups—typically low status groups—who are the ones evaluated by AI. Yet, if human creations are perceived as superior to those of AI, according to the machine penalty, then some valuable human creations—art, advice, products—will be out of reach for some people. Like more immediate consequences of the machine penalty, the key to whether it has beneficial or detrimental consequences depends on whether it is reducing bad outcomes from the AI or stifling good ones.

12.1.3 Harm and Suffering

When autonomous vehicles and precision AI medicine and healthcare serve to reduce human suffering, that certainly is a huge benefit to society. The technological capacities suggest there is a potential for those benefits, especially considering that vehicle accidents and medical blunders *should* be much less common for AI, assuming they are correctly programmed and calibrated. Yet for harm and suffering, reluctance to allow machines to function in these domains has the potential to derail this important consequence.

As we saw in Chap. 9, people are often reluctant to accept AI technologies in domains that can cause harm, especially when those technologies could directly affect them, are less controllable, and have predetermined decisions about how to mitigate harm. Like a social dilemma, people's orientation toward maximizing their own personal safety in their vehicle may serve to actually undermine their safety by making the roads collectively less safe for everyone (Bonnefon et al., 2016).

Adding to this reluctance is a machine penalty of less blame toward autonomous vehicles when they are involved in accidents and human harm—the responsibility gap. Such a combination produces a situation wherein society is slow to adopt the technology that minimizes harm but is also slow to demand that it be fixed when it does lead to harm. Regulation and standards are one way to help dig us out of this dilemma. Industrial factories were once places of incredible suffering, and the suffering was reduced predominantly through laws and regulations. In early factories, the limited automation required humans to be part of their process, sort of like cogs in the machine, whereas current AI usually doesn't expose people to more danger but instead mitigates that danger.

In fact, automation can help move humans out of the most dangerous positions, which is why perceiving AI as having diminished ability to take those jobs tends to work against reducing overall harm. As AI and robotics technology improves, replacing humans with machines in dangerous situations should ultimately reduce harm.

But this depends on the technology's actual ability as well as the ethical issue of what harm outcomes are most vital. For example, a military robot may make a precision strike or defuse a bomb, taking a role previously occupied by human soldiers, thus protecting them from that added harm. But a less sophisticated version of that technology could result in civilian deaths if it fails to adequately respond to a real environment. Penalizing a machine that saves soldiers' lives because it can harm civilian lives may slow down the progress needed to improve the technology to minimize harm to both. Obviously, improved technology in the wrong hands or used improperly can also lead to more harm, such as robotic weapons purposely used to make precision strikes against civilians.

12.2 Consequences for Businesses

12.2.1 Undermining AI Business Solutions

Organizations of all flavors adopt AI to speed up processes, make their business more efficient, and increase their innovation, reach, or influence. Yet, the use of AI itself may have a number of consequences beyond the effectiveness and immediate impact. Clearly, if AI perpetuates unfair evaluations, fails to mitigate harm, or has other negative side effects, they hurt the organization's functionality and image.

An organization's employees may diminish the outcomes of an AI or strategically disregard it, undermining the company's purpose for implementing the AI system. For example, radiologists who ignore superior diagnoses from AI not only produce worse results but also are wasting the organization's resources by not utilizing the AI's recommendation (Agarwal et al., 2023; Bresnick, 2023). In contrast, endoscopists combined their own insights with the AI's diagnoses into solutions superior to both the human-only and AI-only solutions (Reverberi et al., 2022).

For customers too, AI systems designed to improve processes can backfire and cause negative perceptions. A classic case might be an automated phone system or chatbot. While sometimes these increase the efficiency of customer interaction as well as saving the company money, poor answers or slow processes easily cause customer frustration (Chhabra et al.). In extreme cases this can lead to people abandoning advanced technologies (Shank et al., 2022). Smart devices, automated technologies, and AI assistants may all be used within organizations to maximize production, but may lead to negative reactions by both customers (Gonçalves et al., 2020) and employees (Hornung & Smolnik, 2022). Ultimately, this can stem from a machine penalty for the same outcomes, or from AI and humans providing different outcomes.

These examples serve as a warning for business leaders who focus only on the acquisition of a new algorithmic process but fail to see the social psychological barriers involved in its integration with their workforce.

12.2.2 A Bad Reputation from Within

As previously discussed, employees who are evaluated by AI are perceived to be lower status, dehumanized, and have negative emotions (Jago & Laurin, 2017; Jago et al., 2024; Lee, 2018), which leaves a bitter taste for the company using the AI evaluations.

There is also a real possibility that a fairness machine penalty could interact with this dehumanization by machines. For example, the leaders of a company may use an AI system for human resource management that is efficient, accurate, and has been scrutinized and audited to confirm that it is fair and unbiased (Guszcza et al., 2018). Yet, those being evaluated for promotion or hiring might have little knowledge of the machine's overall fairness, highlighting the need for organizations to communicate about the accuracy of AI systems.

An individual receiving a negative evaluation by a machine would likely view it as less fair than a human evaluation, based on the belief that the machine hasn't taken their unique qualifications into account (Lavanchy et al., 2023). Yet they might simultaneously feel dehumanized for being judged by a machine in the first place. These processes could easily feed on each other. It's similar to the ramped-up version of calling a company to deal with a problem, and then experiencing additional delays by having to deal with navigating their automated phone system.

Thus, an organization's internal reputation may sour over time even when using a fair and effective AI, when employees still perceive it as less fair. They'll be especially likely to perceive the organization as less fair if it uses AI for big human resources decisions like promotions and layoffs, leading the employees to report less commitment to the organization (Newman et al., 2020). Again, the human factors of the machine penalty may end up being as critical to an organization as the automation and efficiency benefits of the AI.

12.3 Consequences for Truth

12.3.1 Deepfakes and AI Manipulation

AI are used to compete with and emulate humans in a number of deceptive and manipulative ways, spreading falsehoods, hallucinating facts, and creating manipulated content that previously would have been difficult or impossible to fake. When photography or videography is forged and manipulated—what we call deepfakes—or when AI's creations are passed off as a human's—which includes phishing, spamming, deceptive bots, and AI-based plagiarism—it should be identified as such and taken for what it is.

Our tendency to diminish AI's creations might be partially due to these unethical uses that can harm people and sow disinformation. For example, we believe news headlines less when they are created by AI compared to those written by humans, and we are more likely to view AI headlines as inaccurate (Longoni et al., 2022). If this machine penalty is applied widely to deceptive AI creations, diminishing it might be appropriate. Yet what is really needed is uncovering that deception and creating systems that keep it from occurring and proliferating.

AI also has the possibility of policing manipulative AI. As new generative AI allows people to create essays, photos, images, and videos, other AI systems can analyze that content to detect whether it is likely AI-generated. In this arms race, AI can therefore be on both sides of the truth. Yet other systems—technological, legal, or structural—can be developed to help fight for accuracy and fidelity.

We've done this before with older manipulative technologies. People are still harassed by internet trolls and tricked by phishing and spoofing emails, but social media profiles, verifications, spam filters, warnings, blocking, and reporting can help police manipulative machines and greatly reduce the chance of our being deceived. In this conflict, the machine penalty can be applied strategically to devalue content, such as being applied to the clickbait recommendation from an unknown website, but not from those sites we find reliable.

Yet sometimes—maybe often—it's not even AI vs humans, but the combined efforts of both to manipulate or game. For example, Michael Smith created many fake bands generating AI music, which he attributed to them. Then he created fake fans—AI accounts that "listened" to these fake band's songs. After years and after making $10 million from royalties on Spotify, Amazon Music, and Apple Music, Mr. Smith has been indicted on charges of fraud (Coleman, 2024).

12.3.2 Cultural Erosion

A central social issue and consequence of truth is how AI capacities for deepfakes and deception actually lead us to devalue everything. When we no longer can believe anything, we diminish and are suspicious of everything, leading to problems wherein people disagree on basic facts, trust in institutions breaks down, companies become more powerful than government, and the public square is replaced by polarized and niche communities. While these have all occurred to some degree, super powerful AI can accelerate us toward fragmented societies where AI power and control dominate (Suleyman, 2023).

Mustafa Suleyman, cofounder of DeepMind and an AI leader previously at Google and now Microsoft, does not just powerfully argue that exponentially powerful AI leads to a breakdown of truth and society but also that there is a way out. Specifically, he suggests ideas of slowing down AI development through human systems. These include technological safety, technology audits, slowing development, ethical designers, aligning business incentives, government reformed regulation, international alliances, a culture of shared failures, and an accountable public (Suleyman, 2023, pp. 274–275). If used well, AI does not have to be the enemy of truth.

In fact, AI can be utilized to help repair damaged social fabric, restore trust, and promote truth. New research shows that the large language model GPT-4 Turbo can be used to reduce conspiracy theory beliefs because the AI is able to understand the specific tenets that uphold the belief and tailor personalized counterevidence based on those tenets. Conversation with this AI reduced belief in the specific conspiracy theory discussed, reduced unrelated conspiratorial beliefs, and decreased behaviors related to believing the conspiracy, with those effects lasting months after the short conversation (Costello et al., 2024).

While these outcomes were not directly compared to humans, meaning a machine penalty couldn't be tested, the capacities for AI to have a positive or negative impact on cultural consensus and truth suggest that a penalty against AI could also have a positive or negative impact. We would prefer people not to diminish the effects of advanced AI that reduces conspiratorial beliefs, while we'd be better off if we penalize situations where AI deception is likely.

12.4 Consequences for Uniqueness

12.4.1 What Makes Humans Unique?

AI typically uses human data as their data for learning how to make new creations and decisions. Aside from allowing the deepfakes we just discussed, this also gives rise to questions about copyright and ownership. But, more critically, this challenges the essence of what makes humans uniquely human.

A machine penalty in creative or intellectual endeavors might be an appropriate response to the essentialist beliefs about human–AI differences. While humans, of course, also copy, amalgamate, and learn from others, there is a sense that we add our own "creative juices" and unique perspectives to our new creations and don't just systematically pull from the instances we've seen.

The tension between our beliefs about AI creativity and human uniqueness is nicely illustrated in a conversation between Lt. Commander Data and Captain Picard in the *Star Trek: The Next Generation* episode "The Ensigns of Command" (Snodgrass & Bole, 1989). Picard compliments the android Data's violin playing saying, "Your performance shows feeling," to which Data quickly responds, reminding the Captain that he has no feelings. Captain Picard partially concedes, saying, "It's hard to believe; your playing is quite beautiful."

Data, in his technical and self-deprecating manner, quickly responds, "Strictly speaking, sir, it is not my playing. It is a precise imitation of the techniques of Jascha Heifetz and Trenka Bron-Ken."

Captain Picard pauses at his desk, nods his head side-to-side ever so slightly in disagreement, and asks, "Is there nothing of Data in what I'm hearing?" Then he begins his supportive rebuttal while getting up from the desk: "You see, you chose the violinists. Heifetz and Bron-Ken have radically different styles, different techniques, and yet," now standing face-to-face with Data, "you combined them, successfully."

Data, looking down in thought, concedes, "I suppose I have learned to be creative, sir…when necessary," to which Picard gives an approving nod and smile (Snodgrass & Bole, 1989). Data and Picard resolve that even without the artist's feeling, art can be beautiful, and that while machine creativity on the surface suggests simple imitation, it is often nuanced and complex. The argument is that combining diverse music styles isn't simply AI repackaging human creations.

Whether AI is being creative, in complex conversation, or expressing love for a human, the machine penalty serves as an ongoing but movable barrier between AI's encroaching capacities and the truly unique human. It may be at times that this barrier is an illusion—Data may have been as genuinely creative as a human musician—but the machine penalty shows that we, like Picard, must decide whether to include machines in the sphere of humanity.

The machine penalty suggests we do try to hold onto humans as uniquely creative (Magni et al., 2024). This is part of a larger effect—called the *AI Effect*—whereby the more AI demonstrate being able to match or exceed human capacities, the more we'll see those capacities as nonessential characteristics of being human (Jacobs et al., 2024; Santoro & Monin, 2023) and define other capacities as most essential (Cha et al., 2020). Therefore, new AI abilities should shift—but not obviate—the machine penalty. Likewise, creativity itself may be redefined if we move it out of the essentially human category. Or alternatively, the value of human creativity will be refined and narrowed to only certain aspects where humans can still dominate.

12.4.2 Valuing Humanness

In the short term and for our individual decisions, the machine penalty isn't enough to keep most people from turning away from humans toward AI for cheaper and quicker creations, decisions, conveniences, advice, and perhaps even relationships. When we consider these as meaningfully different outcomes (*Similar Outcomes Condition*) and when AI is superior, we preference it. Therefore, If we value it aesthetically, if it's worth some money, if it serves a purpose, or if it's faster—even if we like it a bit less—many will concede the conveniences are worth it. That is, even with a penalty against it in some cases, AI can fulfill many demands humans have for art, advice, conversations, and products, even establishing a low barrier for new demand.

Yet, the fulfillment of this demand and the creation of new demand should and may coexist with a demand that values the human. In principle, AI art as a whole might highlight the unique aspects of human art, and AI relationships might reveal exactly why human relationships are so valuable (Roose, 2024; Waytz, 2019). In the long run, the machine penalty could mean that human-made products may increase in value, similar to how handcrafted products have a higher market value than similar factory products (Roose, 2022). Faced with a glut of AI art, the scarcity of human art could increase in value.

Likewise, AI advice, relationships, and other conveniences can lead to backlash in which humanity and human interaction and connection are supreme. AI can make us feel heard better than humans can because in this situation their responses are more holistic, clear, and supportive than humans. Yet that superior behavior is diminished when people know the response is from an AI (Yin et al., 2024). In other words, we value a human response beyond the response itself—because it's human.

12.5 Consequences for Relationships

12.5.1 Altering Expectations for Human Relationships

AI has the potential long-term effect of diluting the quality of human relationships. This could happen if AI can deliver on some aspects of a relationship, such as being responsive communicators, but are perceived to inadequately provide in others, such as emotional support, physical intimacy, passion, and long-term responsiveness (Chen et al., 2025; Koike & Loughnan, 2021; Song et al., 2022).

If people recognize these differences and appropriately apply a machine penalty to these relationships, they might accurately understand AI companionship for what it is and what it can and cannot provide. Because the machine penalty rests on what human comparison we make (*AI vs Human Condition*), AI relationships may accurately be viewed as superior to some human relationships, at least in certain aspects. Certainly, even a subpar AI relationship may be better than an abusive human one.

Yet, some have argued that displacing human relationships in the short term can reduce reliance on and expectations for human relationships. Our inability to connect to other humans may have a negative effect on society in the long term (Turkle, 2011). Therefore, again, the machine penalty might be one mechanism by which we as a society socially calibrate to the situations where AI relationships might be most beneficial, while also recognizing the value of human connections.

I've heard an analogy for AI relationships—and potentially AI art, personal advice, and other areas—being compared to junk food, with human relationships being compared to more nourishing food. Yes, junk food tastes good and does satiate our appetite for a while, but, in the long term, our bodies and minds respond better to receiving a balanced and nutritious diet. Therefore, this analogy suggests that AI relationships in moderation might be fine or even beneficial, but you'll be better off with a solid mixture of good human relationships.

Research as of yet cannot support or refute this analogy, as the effect of long-term AI relationships is still largely unknown. I suspect that, like other technologies, such as social media, there may be both positive and negative social consequences (Hancock et al., 2022; Huang, 2017; Saiphoo et al., 2020; Twenge, 2017). But on an individual level, AI relationships may vary a good deal based on one's use, expectations, social connections, temperament, and personality.

12.5.2 AI Companions as Relationship Training

Another possibility for relationships is that people will use AI relationships to help learn how to function better in human relationships. While there are countless self-help books, sometimes practice—and the lack of practice—maintains the haves and have-nots of relationships. In high school and college, those who are attractive, high-status, and socially skilled tend to date more, and those dating experiences often distinguish them further from those who don't date. If set up properly, AI-relationships-for-training could benefit many people, including individuals experiencing the "loneliness epidemic," those with social disorders, those who abuse their partners, and even *incels*—a group of misogynistic men self-named with the portmanteau of "involuntary celibate" who denigrate, objectify, and blame women for their situation.

Whether AI relationships for training will break that cycle depends on the individual but also on the types of AI developed and used. An artificial partner who is programmed to fall in love and who has features that can be altered at whim does not set up expectations for healthy human–human relationships. In fact, AI relationships may set unrealistic expectations for behavior and interaction, in the same way that cultural values of attractiveness, the beauty-industrial complex, and pornography set up unrealistic expectations for beauty.

These relationships could have unrealistic and negative effects. For example, people consider it more acceptable to sexually assault a sex robot (Grigoreva et al., 2024), and people who anthropomorphize female sex robots show increasing hostility toward women (Desbuleux & Fuss, 2023). Yet technologies used to learn, train, and practice, in theory, could help train humans in behaviors to develop and maintain healthy human relationships.

Relational and social AI may be especially beneficial for teaching about platonic friendship, especially to those who have difficulties with social relationships, such as children on the autism spectrum. Here, an AI or robot companion can function like a game that teaches real skills even without having the real consequences. However, the difficulty becomes in distinguishing between real and the practice when the practice seems very real (Elder, 2017).

12.6 The Positive and Negative Consequences of the Machine Penalty

One underlying connection across all these social consequences is that the machine penalty does not have a singular consequence or even one set of good or bad consequences.

At this point, it would be ideal to be able to claim, *Here's the big problem with the machine penalty*, or *Here's what happens if we don't change*. But the machine penalty's consequences are more nuanced. Since we tend to diminish similar outcomes from AI instead of humans across situations, the consequences depend on what outcomes are diminished (*Relevant Outcomes Condition*), whether they are desirable, and other aspects of the situation (*Situational Effect Condition*). If the outcomes are truly the same in all meaningful ways (*Similar Outcomes Condition*), then the penalty is a misjudgment and cognitive error. Yet, we often can't fully compare every outcome and sometimes don't completely know the full outcomes, meaning that the machine penalty is a heuristic which may lead to better or worse decisions.

In short, and like the immediate consequences, we might consider the major difference in downstream consequences based on the positivity or negativity of the outcome. Are AI, compared to humans, leading to better decisions, with less harm and bias, that help us out only when we need help, make us feel good, and positively impact our health and future? If so, the penalty stifles that benefit. Are AI, compared to humans, leading to poorer decisions that are potentially unfair, harming others or ourselves, reducing autonomy, leaving us living in worse situations, and leading to less fulfilling lives? If so, diminishing these AI outcomes is not only appropriate but also desirable. The penalty might be the brakes that help us slow down and not default to AI when there are richer human options available.

Adding to the nuance, many times, we just don't know if the AI's outcome is positive or negative. If we did, we could probably be a lot more strategic about it. Or we might see some positivity mixed in with some negativity. An AI relationship might satisfy us today, but a human relationship might satisfy us beyond today. We may want to drive ourselves, as we perceive such a choice as maximizing our personal safety, even though letting machines drive us might actually promote more general road safety. Given the nuance and specific situations, how do we respond to the machine penalty and its consequences as individuals and society? There's not a simple response, but in the next chapter, I pivot to giving some high-level recommendations as takeaways from these ideas and research.

References

Agarwal, N., Moehring, A., Rajpurkar, P., & Salz, T. (2023). *Combining human expertise with artificial intelligence: Experimental evidence from radiology* (Working Paper 31422). National Bureau of Economic Research.

Bonnefon, J.-F., Shariff, A., & Rahwan, I. (2016). The social dilemma of autonomous vehicles. *Science, 352*(6293), 1573–1576. https://doi.org/10.1126/science.aaf2654

Bresnick, J. (2023). Does AI matter if human clinicians ignore it? *DHInsights*. https://www.dhinsights.org/news/does-ai-matter-if-human-clinicians-ignore-it

Cha, Y.-J., Baek, S., Ahn, G., Lee, H., Lee, B., Shin, J.-e., & Jang, D. (2020). Compensating for the loss of human distinctiveness: The use of social creativity under Human–Machine comparisons. *Computers in Human Behavior, 103*, 80–90. https://doi.org/10.1016/j.chb.2019.08.027

Chen, Q., Jing, Y., Gong, Y., & Tan, J. (2025). Will users fall in love with ChatGPT? A perspective from the triangular theory of love. *Journal of Business Research, 186*, 114982. https://doi.org/10.1016/j.jbusres.2024.114982

Coleman, M. (2024). The bands and the fans were fake. The $10 million was real. *The New York Times*. https://www.nytimes.com/2024/09/05/nyregion/nc-man-charged-ai-fake-music.html?smid=nytcore-android-share

Costello, T. H., Pennycook, G., & Rand, D. G. (2024). Durably reducing conspiracy beliefs through dialogues with AI. *Science, 385*(6714), eadq1814. https://doi.org/10.1126/science.adq1814

Dang, J., & Liu, L. (2024). Extended artificial intelligence aversion: People deny humanness to artificial intelligence users. *Journal of Personality and Social Psychology*. https://doi.org/10.1037/pspi0000480

Desbuleux, J. C., & Fuss, J. (2023). Is the anthropomorphization of sex dolls associated with objectification and hostility toward women? A mixed method study among doll users. *The Journal of Sex Research, 60*(2), 206–220. https://doi.org/10.1080/00224499.2022.2103071

Efendić, E., Van de Calseyde, P. P., Bahník, Š., & Vranka, M. A. (2023). Taking algorithmic (vs. human) advice reveals different goals to others. *International Journal of Human–Computer Interaction, 40*(1), 45–54. https://doi.org/10.1080/10447318.2023.2210886

Elder, A. M. (2017). *Friendship, robots, and social media: False friends and second selves*. Routledge.

Gonçalves, L., Patrício, L., Grenha Teixeira, J., & Wünderlich, N. V. (2020). Understanding the customer experience with smart services. *Journal of Service Management, 31*(4), 723–744. https://doi.org/10.1108/JOSM-11-2019-0349

Grigoreva, A. D., Rottman, J., & Tasimi, A. (2024). When does "no" mean no? Insights from sex robots. *Cognition, 244*, 105687. https://doi.org/10.1016/j.cognition.2023.105687

Guszcza, J., Rahwan, I., Bible, W., Cebrian, M., & Katyal, V. (2018). Why we need to audit algorithms. *Harvard Business Review*. https://hbr.org/2018/11/why-we-need-to-audit-algorithms

Hancock, J. T., Liu, S. X., Luo, M., & Mieczkowski, H. (2022). Social media and psychological well-being. In *The psychology of technology: Social science research in the age of Big Data* (pp. 195–238). American Psychological Association. https://doi.org/10.1037/0000290-007

Hohenstein, J., Kizilcec, R. F., DiFranzo, D., Aghajari, Z., Mieczkowski, H., Levy, K., et al. (2023). Artificial intelligence in communication impacts language and social relationships. *Scientific Reports, 13*(1), 5487. https://doi.org/10.1038/s41598-023-30938-9

Hornung, O., & Smolnik, S. (2022). AI invading the workplace: Negative emotions towards the organizational use of personal virtual assistants. *Electronic Markets, 32*(1), 123–138. https://doi.org/10.1007/s12525-021-00493-0

Huang, C. (2017). Time spent on social network sites and psychological well-being: A meta-analysis. *Cyberpsychology, Behavior, and Social Networking, 20*(6), 346–354. https://doi.org/10.1089/cyber.2016.0758

Jacobs, O. L., Pazhoohi, F., & Kingstone, A. (2024). Large language models have divergent effects on self-perceptions of mind and the attributes considered uniquely human. *Consciousness and Cognition, 124*, 103733. https://doi.org/10.1016/j.concog.2024.103733

Jago, A. S., & Laurin, K. (2017). *Algorithms and dehumanization in hiring*. Academy of Management Atlanta, GA.

Jago, A. S., Raveendhran, R., Fast, N., & Gratch, J. (2024). Algorithmic management diminishes status: An unintended consequence of using machines to perform social roles. *Journal of Experimental Social Psychology, 110*, 104553. https://doi.org/10.1016/j.jesp.2023.104553

Kislev, E. (2022). *Relationships 5.0: How AI, VR, and robots will reshape our emotional lives*. Oxford University Press.

Koike, M., & Loughnan, S. (2021). Virtual relationships: Anthropomorphism in the digital age. *Social and Personality Psychology Compass, 15*(6), e12603. https://doi.org/10.1111/spc3.12603

Lavanchy, M., Reichert, P., Narayanan, J., & Savani, K. (2023). Applicants' fairness perceptions of algorithm-driven hiring procedures. *Journal of Business Ethics, 188*(1), 125–150. https://doi.org/10.1007/s10551-022-05320-w

Lee, M. K. (2018). Understanding perception of algorithmic decisions: Fairness, trust, and emotion in response to algorithmic management. *Big Data & Society, 5*(1), 2053951718756684. https://doi.org/10.1177/2053951718756684

Longoni, C., Fradkin, A., Cian, L., & Pennycook, G. (2022). *News from generative artificial intelligence is believed less.* Proceedings of the 2022 ACM Conference on Fairness, Accountability, and Transparency, Seoul, Republic of Korea. https://doi.org/10.1145/3531146.3533077

Magni, F., Park, J., & Chao, M. M. (2024). Humans as creativity gatekeepers: Are we biased against AI creativity? *Journal of Business and Psychology, 39*(3), 643–656. https://doi.org/10.1007/s10869-023-09910-x

Newman, D. T., Fast, N. J., & Harmon, D. J. (2020). When eliminating bias isn't fair: Algorithmic reductionism and procedural justice in human resource decisions. *Organizational Behavior and Human Decision Processes, 160*, 149–167. https://doi.org/10.1016/j.obhdp.2020.03.008

Reverberi, C., Rigon, T., Solari, A., Hassan, C., Cherubini, P., GI Genius CADx Study Group, & Cherubini, A. (2022). Experimental evidence of effective human–AI collaboration in medical decision-making. *Scientific Reports, 12*(1), 14952. https://doi.org/10.1038/s41598-022-18751-2

Roose, K. (2022). *Futureproof: 9 rules for humans in the age of automation.* Random House Trade Paperbacks.

Roose, K. (2024). Meet my A.I. friends. *The New York Times*. https://www.nytimes.com/2024/05/09/technology/meet-my-ai-friends.html

Saiphoo, A. N., Halevi, L. D., & Vahedi, Z. (2020). Social networking site use and self-esteem: A meta-analytic review. *Personality and Individual Differences, 153*, 109639. https://doi.org/10.1016/j.paid.2019.109639

Santoro, E., & Monin, B. (2023). The AI Effect: People rate distinctively human attributes as more essential to being human after learning about artificial intelligence advances. *Journal of Experimental Social Psychology, 107*, 104464. https://doi.org/10.1016/j.jesp.2023.104464

Shank, D. B., Wright, D., Nasrin, S., & White, M. (2022). Discontinuance and restricted acceptance to reduce worry after unwanted incidents with smart home technology. *International Journal of Human–Computer Interaction, 39*(14), 2771–2784. https://doi.org/10.1080/10447318.2022.2085406

Snodgrass, M. M. W., & Bole, C. D. (1989). The ensigns of command (Season 3; Episode 2) [TV series episode]. In G. Roddenberry, R. Berman, & M. E. P. Pillar (Eds.), *Star Trek: The next generation*. Broadcast Syndication: Paramount.

Song, X., Xu, B., & Zhao, Z. (2022). Can people experience romantic love for artificial intelligence? An empirical study of intelligent assistants. *Information & Management, 59*(2), 103595. https://doi.org/10.1016/j.im.2022.103595

Suleyman, M. (2023). *The coming wave: Technology, power, and the twenty-first century's greatest dilemma*. Crown.

Turkle, S. (2011). *Alone together: Why we expect more from technology and less from each other*. Basic Books.

Twenge, J. M. (2017). *IGen: Why today's super-connected kids are growing up less rebellious, more tolerant, less happy – And completely unprepared for adulthood – And what that means for the rest of us*. Simon and Schuster.

Waytz, A. (2019). *The power of human: How our shared humanity can help us create a better world*. WW Norton & Company.

Yin, Y., Jia, N., & Wakslak, C. J. (2024). AI can help people feel heard, but an AI label diminishes this impact. *Proceedings of the National Academy of Sciences, 121*(14), e2319112121. https://doi.org/10.1073/pnas.2319112121

13

Responding to the Machine Penalty

The machine penalty happens. But what does that mean for me? Or what does that mean personally in my role at work, in relationships with other people, in my hobbies and interests, and in daily life? What does the penalty suggest about policies, laws, or other rules? In this chapter, I have seven responses—general high-level takeaways—for us. Whereas the last chapter was a bit speculative as I projected consequences out to society more generally, this chapter will be even more so, as I go beyond the research conclusions to suggest how we might live differently based on them.

13.1 Response Based on Performance Quality

Couching AI's outcomes as positive or negative at the end of the last chapter, I suggested the major consequences of the machine penalty were in the opposite direction as those outcomes. That is, the machine penalty has a negative impact if it's imposed on a valued, positive outcome, but it has a positive impact if it's imposed on a damaging, negative outcome.

D. B. Shank, *The Machine Penalty*, https://doi.org/10.1007/978-3-031-86061-4_13

Consequently, if we know an AI's capacity to help or be correct, or at least how it compares to humans, we can easily optimize and penalize appropriately and even strategically.

If we don't know the actual performance quality, a little research into an AI's expected outcomes can be helpful. We might not learn definitively whether it will be correct, but we could still get a sense how likely it will be correct or the general quality of its output. Some of this is related to AI literacy—our understanding of how AI algorithms work—but it's also based on finding out how much a model has been tested or audited, what data it was trained on, how it was trained, and other critical information. Some AI systems have information in the form of a model card or a frequent questions page. Some are in Beta testing, while others have been Beta tested. Some have been vetted and recommended by experts in the field.

Consider the generative language-based AI that gives search result summaries or can be queried independently. Knowing how they function and what sort of material they are likely to output helps determine—based on your current situation and desired outcomes—how much to rely on them. We know they give broad summaries based on the most common material they can find online, but when queried more directly, they are more likely to hallucinate and fabricate information. In this case, we should rely on them and not penalize them when we want general information or are checking an uncontested fact. It's a sped-up way to find general information, without digging through websites or considering the myriad of ways people explain the same idea. But when we need details that drill down to specifics, we should penalize AI in favor of our human sources. In this case, we should look up information on websites that other humans made, even if we use the language AI as a starting place.

Therefore, this first recommendation may feel like common sense, but it's worth stating because it's not always easy to put into practice.

Performance Recommendation: Embrace the machine penalty when AI are likely to have inferior performance, but resist the machine penalty when AI are likely to have superior performance.

13.2 Response Based on the Domain

Continuing the previous argument, we should consider how this first recommendation might change based on the domain of operation. The situational causes of the machine penalty show that we tend to undervalue AI in some domains in which they actually perform quite well. We should, therefore, be careful as we navigate using AI in these domains. If an AI is trained on a lot of high-quality data about people's preferences in an area, then it will likely perform well even for personal decisions, subjective issues, aesthetic creations, or in dealing with human life and health.

A good example of this domain-specificity is medical advice. If WebMD or the Mayo Clinic release an AI trained on real medical data, which has been vetted and audited by physicians, then diminishing or refusing its diagnosis could be a serious mistake. Yet, asking ChatGPT for the same diagnosis is more problematic. Here's why: ChatGPT is a great and powerful AI tool, but it's a general language tool, trained on human language data in general and accessing human language data in general. Thus, accepting a diagnosis from it could be akin to getting advice from a sketchy internet post.

Now ChatGPT is not likely to give you the worst internet advice, but it's possible it would; there are no constraints preventing it from doing so. The same could be said about your friend with no medical background, who swears about a home remedy. Hopefully, you would look that up.

Anne, the administrator in my academic department, has given me recommendations both for a laundry soap to reduce poison ivy and which specialty latte to get at Giddy Goat or CoffeeSmiths, our local coffee shops. When she recommended the soap, I didn't use it until reviewing how it affected people. Yet when she recommends the coffee, I go out and try one without a second opinion.

The difference has nothing to do with Anne's expertise—she's neither a dermatologist nor a barista, yet she has both experience in getting rid of poison ivy and good taste in specialty coffees. The difference between the two relates to the quality and impact of the outcomes in each domain.

A bad cup of coffee is a minor inconvenience, whereas an ongoing skin rash has higher importance. Of course, Anne gave great advice in both cases! The point is that we should have a higher bar for things that have a larger impact on our health and wellbeing, and thus be more demanding about the quality and characteristics of the AI.

Similarly, we should penalize AI that are not domain-specific experts in areas that are critical to our health, have ethical ramifications, or generally have more importance to us. Yet, this should not be targeted to all AI, as some may be trained and provide superior results. The simple conclusion is to exert more scrutiny and review in critical domains. Specifically,

> *Domain Recommendation: Carefully consider the quality and domain-specificity of AI, especially for areas of importance, personal wellbeing, and moral decisions.*

13.3 Response Based on Your Interests

Domains not only vary in impact and the need for quality control but also intersect generally with human interests and personally with our interests. We may believe that both a movie recommendation AI and an autonomous vehicle potentially have similar or even superior performance to humans, but there are significant differences between the arenas of these AI. In line with the *Domain Recommendation*, the most obvious one is that the vehicle is responsible for our physical safety, whereas the movie, at most, may influence our emotions and minds or take up some time.

But watching a movie is also a more personal experience, one that we do for our own gratification. In fact, AI only recommends movies; it would be odd to decide to let an AI watch a movie for you. Unless the movie is a task for you—say, for a film class—the point of watching a movie is the experience. That's the primary outcome. In contrast, if the primary outcome is getting to one's destination safely and efficiently, you may willingly cede driving to a machine that can deliver. But, if you want to take a 1920s-style scenic drive, you want to test out the new suspension, or you really like to drive in order to relax, then that changes the relevant outcomes (*Relevant Outcomes Condition*), which makes all the difference.

Keeping the autonomous vehicle and movie AI examples in mind, I'm asking you to consider how AI challenges, supplements, fulfills, supersedes, or undermines your personal desires and goals as it relates to the *Domain Recommendation.* Just because it's important for your health and well-being, doesn't mean it's personally meaningful.

I've heard this sort of recommendation applied, in a non-AI context, to house and yard work. Whether to hire someone to mow your lawn should be based on not only your income, time, lawn size, equipment, and upkeep, but also your intrinsic desire or lack of desire to mow your lawn. I find mowing the lawn a nice exercise on a pleasant day, whereas I'd rather someone else clean the gutters. Cooking is another great example, as many people find it an unnecessary chore, while others see it as their passion and pastime. Understanding your interests is the first step to deciding whether to outsource a task. We should not outsource to AI just because they do a good job, when the real issue isn't quality as much as our own desires. Therefore,

> *Meaningful Recommendation: Outsource to an AI when it's not personally meaningful, but don't outsource to an AI when it's personally meaningful.*

13.4 Response Based on Benefit to Others

AI decisions and actions have effects directly on others—such as when AI evaluates people—and downstream effects as AI systems can spread or reduce inequality, displace jobs, or enhance people's lives. Sometimes technological progress—AI included—seems like an inevitable force that we are powerless to stop (Kurzweil, 2024). At a macro level, this might be true, but choices we make now can affect the future of AI (Suleyman, 2023), and certainly affect how we live our lives (Roose, 2022) and how it affects others (Waytz, 2019).

While not everyone can audit AI systems, we can oppose ones that bring clear negative effects to other people, even if they bring us potential advantages. Similar to the previous argument for the *Domain Recommendation* suggesting more scrutiny for ethical, health, and well-being decisions, when an AI is used to evaluate people, distribute resources, or even propagate information, this is an opportunity for us to pause and consider the potential effects.

We should ask ourselves: Will this machine likely cause more harm or less harm? Will some groups be systematically and unfairly disadvantaged by use of this AI? Will this AI spread misinformation or ideas that would hurt people? And, as much as we are able, we should decide to promote and support AI that align with these values, while boycotting and supporting changes for AI that don't. Depending on how we answer those questions in a situation, we may find we side with or against the AI. Subsequently, we would be aligned against penalizing beneficial AIs or for penalizing detrimental ones.

> *Human Benefit Recommendation: Double down on penalizing machines when they can perpetuate unfairness or harm toward people, but resist penalizing machines when they benefit people and society.*

13.5 Response Based on the Value of Human

Sometimes AI can benefit or harm others, but other times, there are not clear upsides or downsides to other people. Another way other people play into our decisions regarding AI is the value that humans may provide to an outcome. Human value would apply directly to artistic creations, but for other products, advice, or services, humans can also add value. Humans are unique, from having vision, style, and idiosyncrasies, to having emotional and motivational reasons for creating and serving.

Dining services illustrate human value. A full-service restaurant is based on the idea that humans anticipate, optimize, and attend to our needs as part of the experience of eating a meal. Yes, robots and AI can—or soon will be able to—handle seating, ordering, cooking, and serving the food. Yet many restaurants have not adopted even clearly cost-saving machines, because they believe humans can offer value to the experience that robots and AI cannot (McCarron, 2024).

First, humans can often react and customize to specific customer's dietary needs from gluten-free to letting you sample a salad dressing before committing to it. Second, this can expand to personal recommendations for food and personal interaction in service. The best waiters and waitresses adapt to their patrons, responding differently to the foodies who want to try everything, the loner who would love to chitchat, and the businessmen who would rather not be interrupted. Third, the value sometimes comes from the relationships formed. We might have a favorite server at a restaurant or a local barista who knows our name and favorite drink. While AI may be able to accomplish these more social tasks, at least for now, the restaurant experience seems to benefit from the human connection that comes from another person adapting to your needs in this social service situation. All of these human interactions add value to a product or outcome that could otherwise be created or delivered by AI.

Best Buy CEO Hubert Joly created human value in selling their products in an Amazon-dominated market. Best Buy's products are mainly technologies, which may seem the furthest from areas like food and art, where human value may be more expected.

Joly realized that people were coming to Best Buy to interact with the products themselves. Beyond the specs, customers were concerned about their personal experience with the technology and would use Best Buy to *showroom*— that is check it out before buying it online. Joly lowered the prices to make sure those coming in the door for that purpose would result in a sale for them, but he also went a step further. He transformed his workforce into a team of personalized technology concierges. Modeled after Best Buy's Geek Squad, this new team helped customers select, understand, try out, and personalize the technology based on their needs (Roose, 2017, 2022).

In short, they flipped from showrooming to showcasing. Ultimately, the value of *humans* has kept them thriving in a world where many other technology-selling box stores have gone under.

In contrast, sometimes there is little to no value added by humans. When this is the case, an AI advantage is appropriate. If you don't want to have pleasantries with a grocery store clerk and there is nothing complicated about your purchase, the human cashier will offer little advantage over the self-checkout. Likewise, for many discrete products, suggestions, or information, there is no clear human element and, therefore, no need to advantage humans. This is the exception that proves the rule, as not everything gains value due to humans. But, if humans really do add value, we could incorporate that value into our decisions. Overall,

Human Value Recommendation: Value human outcomes for what is uniquely human about them.

13.6 Response Based on AI Not Being Human

Continuing with the grocery store scenario, we may not find a substantial difference between using a self-checkout—which is hardly a proper AI—and checking out with the human cashier. Yet, even if human value doesn't add to this outcome, and even if we don't desire humans to provide a different outcome, there is a critical difference. The human cashier is human. AI are not.

In Chap. 2, I argued that discrimination can have different meanings we might apply to AI. One way to conceptualize that is: that it's appropriate to discriminate, as in differentiating based on qualities, among either humans or AI.

But prejudicial discrimination is worlds different for humans compared to AI. I don't see anything inherently wrong with prejudicial discrimination toward AI—at least AI as they exist now—whereas I believe prejudicial discrimination toward other people is clearly wrong. There is a certain intrinsic value to being human, whether you think of this in terms of unalienable rights, a spiritual soul, or simply your own first-person experience of being sentient. And I believe that this puts humans in a special category for respect and decency. Now, I'm not for animal abuse, defacing property, destroying the environment, or being vulgar toward AI, but directing those vices toward other humans makes them ethically worse, due to humans having additional intrinsic value. Yet, even cruelty toward nonhumans may promote cruelty toward humans (Coghlan et al., 2019), and thereby have indirect moral consequences for humans.

This pro-human view is not an uncontroversial and uncontested opinion. Many people—depending on their religious beliefs or views about animals' or robots' capacity—come to different conclusions about their levels of mind and essential nature, thereby believing they deserve moral consideration (Pauketat & Anthis, 2022) or have a soul. For example, nearly 20% of people believe current AI is sentient (Pauketat et al., 2023), and sentience—and potentially other criteria—suggest that future AI may deserve moral consideration (Ladak, 2024).

But even if you elevate nonhumans, you probably realize the intrinsic value of humans. The key here is to not just know this intellectually, but to act on this belief: both realizing and respecting the value of people. When our Alexa or Siri gives us a response, saying "thank you" may be our natural response—and good practice—but ultimately serves no benefit to the smart assistant. It is an AI, a program, and—as AI exists now—ultimately a thing. In contrast, if a friend named Alexa or Siri gives us that same response, "thank you" represents not only adhering to polite human norms but also acknowledging the value and humanity of that person.

We often do this for friends, but—as in the cashier example—we are less likely to do this with strangers in service, business, and mediated environments. Since the industrial revolution, people have been used as part of larger systems, including the *McDonaldization* of our society's service industry, wherein human service is quantified and rationalized to hide human value and uniqueness (Ritzer, [1993] 2004). This means that humans are the cogs in systems that function like machines. Remember that cashiers, waiters, artists, and those who read our online posts are valuable people even—and maybe especially—if we don't know them or disagree with them.

Therefore, the Bible's Golden Rule, "Do unto others as you would have them do unto you," is a great maxim for interacting with other people, but it doesn't have to apply to AI. If we end up enhancing our respect toward and treatment of other humans—even if we diminish it toward AIs—I believe the world would benefit, as many problems are from humans treating humans poorly. I think my recommendation here is best summarized in the title of a minimalist book on how things can detract from important human relationships. It's titled *Love People, Use Things: Because the Opposite Never Works* (Millburn & Nicodemus, 2021). Despite all their capabilities and all our social reactions to them, AI—for now—are still things, but humans are people. Therefore,

Loving People Recommendation: Love people, use things.

13.7 Response Based on Being Human in an AI World

This is not a book about automation or AI taking our jobs, but that is a legitimate concern for many people (Dahlin, 2019). In many ways, a human and an AI competing for the same job reflects an ultimate application of the machine penalty. Usually, however, from the perspective of the management considering automation, the AI replacing us produces a superior set of outcomes, especially if it has the mind and capacities that fit with that job (Yam et al., 2025). Furthermore, people have greater approval for and find it more morally justifiable to replace people with machines, compared to replacing them with other groups of people (Hidalgo et al., 2021). Yet, if you're the one being replaced, then you would probably diminish the machine, based on the personal and controllable machine penalty causes.

But in a world where we could be devalued and dismissed in place of machines, how do we retain the humanity that has been so important in the social consequences and other responses to the machine penalty? Here, I want to pull from Kevin Roose's book, *Futureproof: 9 Rules for Humans in the Age of Automation*, wherein he explains, in detail, the uneven and contextual factors in technological automation, and then spends the bulk of the book providing a personal playbook for being human in an automated world (Roose, 2022).

His rules fall into a few categories. Several are about doubling down on what makes us human or what we as humans can do especially well compared to AI, similar to my *Human Value Recommendation*. This includes embracing creative tasks, rare tasks, and socially human tasks. And our skills for these only improve the more we engage with humanity and with the humanities. Likewise, there are many important aspects of human connection like building up healthy relationships and networking, as they are powerful because they are based on human–human links, and therefore not easily replaced by AI.

A second category is to act with agency toward your devices and machines. You, not your phone or email, should control your day. And you should let your passions, not your metrics, drive you. Related to the second category—and to several of my recommendations—is a third category, which is to decide how to use AI, which means you sometimes resist AI. Use AI when it can be truly beneficial, but also critically evaluate each of the technologies you are using (Roose, 2022). A machine penalty may be an excellent response in-line with thinking through the value of machines in how they influence your life. All of these suggest personal alignment, so that you are not competing with machines in situations where they will win. Specifically, considering these suggestions,

Futureproof Recommendation: Futureproof yourself against AI and automation.

13.8 Summary of Recommendations

Table 13.1 summarizes my seven recommended responses to the machine penalty. Hopefully, this provides us insights on what we might do in response to the penalty. It's sometimes difficult to predict longer-term consequences and know how to respond, especially as AI continues to improve and change. We don't know the future, but next, I conclude by considering how the machine penalty might apply or change in the future, when AI will be very different than they are today.

Table 13.1 Seven recommended responses to the machine penalty

Recommendation	Type
Performance Recommendation	Embrace the machine penalty when AI are likely to have inferior performance, but resist the machine penalty when AI are likely to have superior performance.
Domain Recommendation	Carefully consider the quality and domain-specificity of AI, especially for areas of importance, personal wellbeing, and moral decisions.
Meaningful Recommendation	Outsource to an AI when it's not personally meaningful, but don't outsource to an AI when it's personally meaningful.
Human Benefit Recommendation	Double down on penalizing machines when they can perpetuate unfairness or harm toward people, but resist penalizing machines when they benefit people and society.
Human Value Recommendation	Value human outcomes for what is uniquely human about them.
Loving People Recommendation	Love people, use things.
Futureproof Recommendation	Futureproof yourself against AI and automation.

References

Coghlan, S., Vetere, F., Waycott, J., & Barbosa Neves, B. (2019). Could social robots make us kinder or crueller to humans and animals? *International Journal of Social Robotics, 11*(5), 741–751. https://doi.org/10.1007/s12369-019-00583-2

Dahlin, E. (2019). Are robots stealing our jobs? *Socius, 5*, 1–14. https://doi.org/10.1177/2378023119846249

Hidalgo, C. A., Orghian, D., Canals, J. A., De Almeida, F., & Martin, N. (2021). *How humans judge machines*. MIT Press.

Kurzweil, R. (2024). *The singularity is nearer: When we merge with AI*. Random House.

Ladak, A. (2024). What would qualify an artificial intelligence for moral standing? *AI and Ethics, 4*(2), 213–228. https://doi.org/10.1007/s43681-023-00260-1

McCarron, M. (2024). Do robots love their customers? Automated restaurants face human issues. *The New York Times*. https://www.nytimes.com/2024/08/13/dining/ai-robots-restaurants.html

Millburn, J. F., & Nicodemus, R. (2021). *Love people, use things: Because the opposite never works*. Celadon Books.
Pauketat, J. V. T., & Anthis, J. R. (2022). Predicting the moral consideration of artificial intelligences. *Computers in Human Behavior, 136*, 107372. https://doi.org/10.1016/j.chb.2022.107372
Pauketat, J. V. T., Ladak, A., & Anthis, J. R. (2023). Artificial Intelligence, Morality, and Sentience (AIMS) survey: 2023 update. *Center for Open Science*. https://doi.org/10.31234/osf.io/9xsav
Ritzer, G. ([1993] 2004). *The mcdonaldization of society* (Rev. ed.). Pine Forge.
Roose, K. (2017). Best buy's secrets for thriving in the Amazon age. *The New York Times*. https://www.nytimes.com/2017/09/18/business/best-buy-amazon.html
Roose, K. (2022). *Futureproof: 9 rules for humans in the age of automation*. Random House Trade Paperbacks.
Suleyman, M. (2023). *The coming wave: Technology, power, and the twenty-first century's greatest dilemma*. Crown.
Waytz, A. (2019). *The power of human: How our shared humanity can help us create a better world*. WW Norton & Company.
Yam, K. C., Eng, A., & Gray, K. (2025). Machine replacement: A mind-role fit perspective. *Annual Review of Organizational Psychology and Organizational Behavior, 12*. https://doi.org/10.1146/annurev-orgpsych-030223-044504

14

The Machine Penalty in the Future

Books about AI can become outdated very quickly. One challenge is to keep updated with the latest research and AI news stories. Between the time I write this and the time you read it, no doubt there will be influential developments in AI and other technologies. However, I'm hopeful this book will have some staying power, because fundamentally it's not about AI. It's about us.

We as humans are the ones who initiate and experience the machine penalty. One-half of the machine penalty's comparison is to a human or humans (*AI vs Human Condition*), and the consequences are consequences for us as the perceiver (Part II), as a society (Chap. 12), and personally (Chap. 13). While we live in a world of rapid advancements in artificial intelligence which fundamentally reshapes our lived experience, our brain structures and psychological motivations are largely unchanged from our ancestors. But our brain's plasticity means that we also psychologically adapt and learn based on culture, experiences, and, in part, technological advances. This psychological makeup suggests that, while AI might be completely different in a few years, we will continue to exhibit

D. B. Shank, *The Machine Penalty*, https://doi.org/10.1007/978-3-031-86061-4_14

similar psychological and social responses. But these responses will be updated as we encounter and form new expectations for future technologies.

To conclude, I would like to consider this theory's ability to be applied to future AI, even though they are, as yet, unknown. No doubt, as AI creeps more into current human domains—and perhaps even help create new domains of interest—there will be new situations to compare machines to humans. In some cases, the increase in comparisons could enable more instances of the machine penalty, whereas, in others, AI may produce such dissimilar outcomes that there might be fewer times to penalize machines. Let's consider some of the theory's fundamentals in relation to how AI developments might play out in the future.

14.1 The *AI Effect* and the Unattainable Human Benchmark

In my *Introduction to Artificial Intelligence* class in graduate school, aside from the formal definitions of artificial intelligence, there was also a tongue-in-cheek definition: artificial intelligence was whatever machines did in the movies. While not the most scholarly and precise way to define AI, there is a sense that not only does fictional storytelling capture potential technological futures, but also that true AI is always somewhere in the future.

Nowadays, it might seem like AI has arrived. There is a proliferation of news articles and books on it, with companies and nations endlessly discussing and investing in it. Or has it arrived? Strong AI, artificial general intelligence, and personal robot companions still seem out of reach, while machines conversing with us, planning agendas, summarizing, creating music, and giving us directions seem too mundane to actually be intelligent.

This always-in-the-future-and-out-of-reach effect is a real phenomenon—a psychological phenomenon, known as the *AI Effect* (Santoro & Monin, 2023). It involves the idea that a computational problem that at one point seems like an important step in artificial intelligence, once solved appears as a simple computer program. Looking forward, it's a breakthrough representing a benchmark of intelligence; while looking backward it was just an inevitable technological milestone (Bailey, 2016; Haenlein & Kaplan, 2019). So, Chess, Go, passing the bar exam, or driving a vehicle seem less like indicators of intelligence now, simply because machines can do them well.

Yet the primary reason for this effect is not about technology or hindsight, but rather making a comparison—and contrast—to ourselves and humanity as a whole (Jacobs et al., 2024). It's less about intelligence and more about whether humans and AI are different, or what essential differences we perceive separate us from them. When AI demonstrates a new ability, we shift that ability out of the human-only camp and into a camp that includes both humans and AI. Whatever is left in the human-only camp is then considered even more essential and critical to being human (Santoro & Monin, 2023).

One reason the machine penalty will likely persist is that it's built on this idea of a human benchmark and our related self-serving bias, whereby we use ourselves as the standard. Whether humans—or biological beings—have intrinsic and immutable differences from AI—or mechanical beings—is a big question tackled by philosophers. However, my argument here is smaller and more focused: we humans are fond of the machine penalty, because it makes us feel superior. Therefore, our self-serving bias—manifested as the *AI Effect*—will lead us to find new human–AI distinctions, and then penalize AI on those, even with AI of the future.

14.2 The Future of the Machine Penalty Causes

In Chap. 6, I laid out five human–AI comparisons as causes of the machine penalty. These perceived differences are unlikely to disappear even with more sophisticated AI and robots, including artificial general intelligence. But arguably, they will change.

For appearance, computer-generated characters and deepfakes can now reliably emulate and pass as humans, suggesting both cases where outcomes will be similar based on appearance and cases where we can't distinguish a machine from a human. But this cause is also based on what sort of appearances are actually implemented. Pixar still prefers clearly animated characters, some far from human, and smart home speakers often have few physical anthropomorphic features by design.

Likewise, identities for AI are based more on consumers' needs and technologists' designs than on the achieved and ascribed statuses that make up human identities. AI are primarily given human identities as signals that they are designed to function in specific ways associated with those identities. Of course, more generalist AI systems may have the ability to subsume many human identities. But unless they choose their own identities, earning them through training and achievement, we are likely to view machine identities quite differently from those of humans.

AI behavior, like appearance, continues to get closer to human capacities. We see this often in new robot designs mimicking and mastering human's physical abilities, like dance or complex hand movements. However, there are growing numbers of AI behaviors differentiating machines from humans, such as ultrafast stock trading and very positively toned written responses. While collectively AI can emulate many behaviors, many machines are specifically relegated to one type, such as autonomous vehicles driving. Similarly, AI typically has standardized behavior, such as smart assistants giving everyone the same answer. Machines with artificial general intelligence will also be less constrained than an AI that operates in a single domain.

Machine minds are more stable across machines and across time, but as the overall level of technological sophistication increases, people revise their expectations for machines, including perceptions of their minds (Pauketat et al., 2023). For example, machines used to not be able to write and produce creative products, but now they compose poetry, write essays in different tones, and make music and artwork. Holistically, this might suggest perceiving more agency as the AI intentionally creates, but also perceiving more experience as the AI reflects more personality, emotions, and memories.

Similarly, AI essence would appear to be a more stable human–AI comparison, but as shown with the *AI Effect*, over time even the essence of AI is susceptible to change (Pauketat et al., 2023). Here, as machines collectively are able to accomplish new feats, we shift the criteria we use to define their nature to ultimately keep them separate from humans.

Situational causes could change over time as well, as AI is developed and used increasingly in all areas of human life. It's reasonable to imagine AI of the future spreading into new domains with related situations wherein the AI's outcomes are personal, important, subjective, or moral and potentially take away our control.

Across all these increased uses of AI, the machine penalty suggests we may elevate human outcomes compared to AI. However, it's also possible that some of the situational causes will weaken if beliefs about machines change. For example, if AI continually demonstrate great capability and reliability for personal or subjective decision-making, our expectations may shift to penalize them less based on those causes. Therefore, these situational causes may be less of universal causes, but instead tied to current norms and related expectations. In that case, these situational causes may be less central to the machine penalty theory in the long term.

14.3 Hybrid Human–AI Teams

One of the biggest future applications of the machine penalty—but also its possible downfall—is human–AI teams and partnerships. Co-creation, hybrid human–machine teams, decision support systems, and other processes where humans and AI work together have been increasing and undoubtedly will be vital in the future. AI can be useful and complementary teammates, as they have abilities to communicate in natural language, know or learn our preferences, and access different sources of data.

From an outsider's perspective, a decision, creation, or other outcome jointly produced will not be clearly attributable to the AI agents or the humans on a team. Yet many of the same causes that shape the machine penalty might affect attributions of the team's behavior. If a human–AI team jointly designs the furnishings and décor for a room, we might assume that a human picked the colors, materials, and feng shui, while an AI optimized product price, quality, and layout efficiency. That is, based on the level of subjectivity of components of the outcome, we might differently attribute contributions to the AI and the humans on the team. In contrast, if we don't know and don't make these assumptions about who did what, the machine penalty would no longer apply (*AI vs Human Condition*).

Yet from a vantage within that team, the machine penalty would directly apply. With the human–AI team, we might assume our human teammate's choice of fabric was superior to the AI's, while we judge the product quality determined by the AI teammate as the superior, because the AI selected it.

In summary, current and future hybrid teams could both complicate and extend the machine penalty.

In Chap. 3, I described three floors of how we perceive machines (Fig. 3.1) including a first floor of seeing them socially, a second floor of seeing them as humanlike, and a third floor—where the machine penalty resides—of comparing them to humans. If I had to propose what a fourth floor would entail, I would surmise that it would involve perceiving teams of AI and humans. As such, on this floor, we would examine how we see humans and AI jointly making decisions or producing outcomes, with one augmenting the performance of the other. Of course, another possibility is that humans are actually augmented with AI.

14.4 Humans Becoming More AI

At some points, technology becomes so intertwined that it becomes integrated as part of one physical self or psychological identity. Being augmented by technology can manifest as becoming a cyborg or transhuman—human's physical integration with technology—but physical integration is just one aspect of psychologically being augmented.

Consider the simplified scenario of physical integration in eyesight correction. One may have glasses, obviously noticeable and easily removable, or contact lenses, less noticeable and removable with just a bit more effort. Those seem like separate physical entities that augment to correct one's vision. But what if the contact lenses were built in and not removable? What about Lasik surgery which reshapes one's cornea? Or what if biochemicals or nanites were added to correct vision? At what level is it physically separate?

There are many levels of physical augmentation for eyesight, but for all of them, we generally consider the final eyesight our own. If you ask someone wearing glasses, "Can you read that sign?" we would be taken aback if they replied, "No, I can't. But I can with my glasses on."

As tool-using entities, we are quick to consider the tool as a part of ourselves. Access to internet searches, for example, makes us believe we know more than we actually do, mistaking the searched information for our own personal understanding (Eliseev & Marsh, 2023; Fisher et al., 2015; Pieschl, 2021). In spite of an internet search being on a device that is not physically connected to our bodies, and in spite of that information not residing in our own brain, we are quick to integrate that tool and its outcome into our view of ourselves. An even simpler example is when someone asks you if you know what time it is. If you're wearing a watch—or have on you another time-telling machine—you often feel like you know, even though it involves consulting a mechanical device.

Thus, something that is *psychologically* part of us does not have to be *physically* part of us. While brain–machine interfaces or other cybernetic enhancements could put computational intelligence into our brain or body physically, we already have artificial intelligence on the phones in our pockets. Psychologically, this means that if we depend on Bard, ChatGPT, or a personalized AI assistant, we'll likely integrate that AI and its abilities into our view of ourselves.

Also, like eyeglasses, it's usually somewhat obvious when we're accessing technology, even though someone on their phone could be texting, searching, or consulting AI. Future generation technology will be more like contacts and be less obvious to outsiders, further collapsing the perception of humans versus hybrid human–AI teams.

The machine penalty may differently apply if, at some point, we assume most humans have AI augmentation available. At that time, we might sometimes compare humans to AI, but sometimes assume "humans" include AI augmentation, like many humans today include corrected eyesight. Perhaps AI will be such an integrated tool that we won't typically even take notice of it.

14.5 AI Becoming More Human

It might sound odd to say that AI is trying to become human. But my final idea is not about a sentient android with a volitional desire to be a human, but about how AI is created to survive in human society.

First, many—if not most—AI are designed to mimic human functions and serve in human roles. One continued trajectory of profit-seeking companies is to copy human functionality, since that allows them to make money by replacing humans and fulfilling human needs.

Second, AI are trained with human data, which allows them to better emulate humans, but also is fundamental to their intelligence. AI needs data almost the way humans need food. When AI are trained recursively on the data they generate, the models degrade and eventually collapse into predicting or producing garbage or identical output. Their output converges to a modal average that's often distinctly nonhuman and nonsensical (Shumailov et al., 2024). Therefore, AI seems fully dependent on humans to continue supplying them with data on how to emulate humans and do basic tasks, not the least of which is adapting and learning in a changing world.

Yet, AI mainly learns from the digital world, which reflects the real world but is not the real world. As they feed output back into the digital world, they create a recursion of garbage, taking as input what they've created as output.

We see this most clearly in stock-market crashes perpetuated by AI. The trading bots exchange with other AI trading bots creating destructive cycles, wherein they respond to each other, not to the real-world companies whose value those stocks represent and the real-world events that may affect them. The same situation could happen in war, where AI weapons respond to other AI weapons in a cycle not related to or even reacting to the constituents of the countries those autonomous weapons represent. AI do not have their own society; they live in ours.

Perhaps the next generation of AI and robots will explore and discover by interacting with their environment, becoming more humanlike as in the children's story *The Wild Robot*. The book—and now movie—tells of the robot Roz who is accidently marooned and activated on an island with no directives other than her preprogrammed instincts and skills.

Through her interaction with the environment including the weather, land, vegetation, and animals—and only later other robots and humans—she learns, grows as an individual, and charts her own path. She learns about overcoming prejudice by adapting to the expectations of the island's animals. She learns about responsibility by caring for a gosling after accidentally killing the gosling's family. She learns about family and longing when that gosling imprints on her as his mother, but eventually must leave her to migrate for the winter. She learns about sacrificial relationships—what we might call love—when she saves the animals of the island from a harsh winter, and they save her from her company's recovery robots (Brown, 2016).

This *Wild Robot*-learning is a decisively different, albeit slower, way for machines to gain intelligence than from being fed a diet of human's digital data. It's exploring food and its culinary delights, not being force-fed. Machine learning techniques, such as unsupervised learning, allow AI to recognize and learn from data with few set parameters or goals. These techniques, combined with autonomous bodies, could provide a foundation for future robots to learn about the environment in the wild.

If we don't give AI and robots independent experiences and they don't develop capacities for independent exploration, I worry the best of AI will still be sophisticated mimicry. Often AI's best answers are reproductions of the best human answers, and their creations are the amalgamation of human creations. As a system, this cuts off avenues where machines can really differentiate themselves from humans, as they are always, ultimately, imitating humans. Thus, I believe there will always be some machine penalties as long as AI are human imitators. Ironically, to really be human, AI shouldn't try so hard to be human.

References

Bailey, K. (2016). Reframing the "AI Effect". *Medium*. https://medium.com/@katherinebailey/reframing-the-ai-effect-c445f87ea98b

Brown, P. (2016). *The Wild Robot* (Vol. 1). Little, Brown Books for Young Readers.

Eliseev, E. D., & Marsh, E. J. (2023). Understanding why searching the internet inflates confidence in explanatory ability. *Applied Cognitive Psychology, 37*(4), 711–720. https://doi.org/10.1002/acp.4058

Fisher, M., Goddu, M. K., & Keil, F. C. (2015). Searching for explanations: How the Internet inflates estimates of internal knowledge. *Journal of Experimental Psychology: General, 144*(3), 674–687. https://doi.org/10.1037/xge0000070

Haenlein, M., & Kaplan, A. (2019). A brief history of artificial intelligence: On the past, present, and future of artificial intelligence. *California Management Review, 61*(4), 5–14. https://doi.org/10.1177/0008125619864925

Jacobs, O. L., Pazhoohi, F., & Kingstone, A. (2024). Large language models have divergent effects on self-perceptions of mind and the attributes considered uniquely human. *Consciousness and Cognition, 124*, 103733. https://doi.org/10.1016/j.concog.2024.103733

Pauketat, J. V. T., Ladak, A., & Anthis, J. R. (2023). Artificial Intelligence, Morality, and Sentience (AIMS) survey: 2023 update. *Center for Open Science*. https://doi.org/10.31234/osf.io/9xsav

Pieschl, S. (2021). Will using the Internet to answer knowledge questions increase users' overestimation of their own ability or performance? *Media Psychology, 24*(1), 109–135. https://doi.org/10.1080/15213269.2019.1668810

Santoro, E., & Monin, B. (2023). The AI Effect: People rate distinctively human attributes as more essential to being human after learning about artificial intelligence advances. *Journal of Experimental Social Psychology, 107*, 104464. https://doi.org/10.1016/j.jesp.2023.104464

Shumailov, I., Shumaylov, Z., Zhao, Y., Papernot, N., Anderson, R., & Gal, Y. (2024). AI models collapse when trained on recursively generated data. *Nature, 631*(8022), 755–759. https://doi.org/10.1038/s41586-024-07566-y

Acknowledgments

I conduct most of my research with students and other faculty, and I teach two classes on how teams work. Therefore, while authoring a book on my own was a departure from that model, I depended on students, collaborators, colleagues, friends, and family throughout the process. They offered advice, support, critiques, and resources, which all led to a better final product than I could manage on my own. So, I would like to thank them all.

First, Michael Peterson helped jumpstart this process by telling me about his experience publishing with Palgrave Macmillan and introducing me to the editors there. Additionally, that publication team including Liam, Kishor, and others provided guidance and answered questions throughout the process.

Second, five student researchers—Eyuel Getahun, Grace Hall, Aiden Pickett, Celia Freed, and Nolan Yee—agreed to be part of my "book team," spending time over the summer searching for the most up-to-date literature. Each focused on a deep-dive literature search for one of the five applied domain chapters in Part II. In addition to helping uncover that literature, they each provided feedback on intermediate drafts of chapters as well as the entire book.

D. B. Shank, *The Machine Penalty*, https://doi.org/10.1007/978-3-031-86061-4

Third, Sarah Hercula, Yochanan Bigman, Jenny Davis, Vahe Permzadian, and David Wright each gave valuable feedback on different aspects of the book. This feedback related to specific questions, to different chapters of the book, to specific technologies, or on framing the book as a whole.

Fourth, four individuals gave me extensive feedback on the entire book, helping me correct issues, reframe arguments, and present a smoother academic, yet accessible, narrative. Nicholas Lewis was gracious not only with connections, corrections, and inquiries but also with feedback on which passages best communicated to a wider readership. Arthur Jago's comments helped me in clearly presenting my arguments and individual research in more accessible ways, while keeping true to the academic literature. Janet Pauketat suggested a number of research studies which I've included, as well as challenged my arguments to include and connect to more philosophic work. Devin Burns was the feedback warrior, generously providing comments from micro-corrections to ideas about the nature of humans and machines. He helped uncover every logical hole and provided counterarguments which greatly sharpened how I presented the theory.

Fifth, my family has been supportive and invaluable through the writing process. Victoria edited a full draft both before and after everyone else did. She and our kids graciously dealt with my extra workdays and long hours as I put this together. I love them and am very grateful for their support.

Finally, I've dedicated this book to my parents and brother, who have always incorporated books, words, reading, and communication as a part of our lives. My brother Nathan is a wordsmith, poet, book lover, book seller, and English professor. Mom is an avid reader, Scrabble player, speaker, illustrator, and proofreader. Dad throughout his career has communicated through spoken and written words, including writing and publishing multiple books. I love them and dedicate this book to them.

References

Agarwal, N., Moehring, A., Rajpurkar, P., & Salz, T. (2023). *Combining human expertise with artificial intelligence: Experimental evidence from radiology* (Working Paper 31422). National Bureau of Economic Research.

Ahn, J., Kim, J., & Sung, Y. (2021). AI-powered recommendations: The roles of perceived similarity and psychological distance on persuasion. *International Journal of Advertising, 40*(8), 1366–1384. https://doi.org/10.1080/02650487.2021.1982529

Alabed, A., Javornik, A., & Gregory-Smith, D. (2022). AI anthropomorphism and its effect on users' self-congruence and self–AI integration: A theoretical framework and research agenda. *Technological Forecasting and Social Change, 182*, 121786. https://doi.org/10.1016/j.techfore.2022.121786

Alimamy, S., & Kuhail, M. A. (2023). I will be with you Alexa! The impact of intelligent virtual assistant's authenticity and personalization on user reusage intentions. *Computers in Human Behavior, 143*, 107711. https://doi.org/10.1016/j.chb.2023.107711

Andrews, E. (2023 [2015]). Who Were the Luddites? *History.com*. https://www.history.com/news/who-were-the-luddites

Angwin, J., Larson, J., Mattu, S., & Kirchner, L. (2016). Machine Bias. *ProPublica*. https://www.propublica.org/article/machine-bias-risk-assessments-in-criminal-sentencing

D. B. Shank, *The Machine Penalty*, https://doi.org/10.1007/978-3-031-86061-4

Ansani, A., Koehler, F., Giombini, L., Hämäläinen, M., Meng, C., Marini, M., & Saarikallio, S. (2024). AI performer bias: Listeners like music less when they think it was performed by an AI. *Open Science Foundation Preprint*. https://osf.io/preprints/osf/3rnve

Appel, M., Izydorczyk, D., Weber, S., Mara, M., & Lischetzke, T. (2020). The uncanny of mind in a machine: Humanoid robots as tools, agents, and experiencers. *Computers in Human Behavior, 102*, 274–286. https://doi.org/10.1016/j.chb.2019.07.031

Arnestad, M. N., Meyers, S., Gray, K., & Bigman, Y. E. (2024). The existence of manual mode increases human blame for AI mistakes. *Cognition, 252*, 105931. https://doi.org/10.1016/j.cognition.2024.105931

Associated Press. (2023). Man 'encouraged' by AI chatbot 'girlfriend' to kill Queen Elizabeth II receives jail sentence. *Euronews*. https://www.euronews.com/next/2023/10/06/man-encouraged-by-an-ai-chatbot-to-assassinate-queen-elizabeth-ii-receives-9-year-prison-s

Augustine, A. A., & Eyssel, F. F. (2024). Motives and risks of self-disclosure to robots versus humans. *ACM Transactions on Human-Robot Interaction*. https://doi.org/10.1145/3700887

Awad, E., Dsouza, S., Kim, R., Schulz, J., Henrich, J., Shariff, A., et al. (2018). The moral machine experiment. *Nature, 563*(7729), 59–64. https://doi.org/10.1038/s41586-018-0637-6

Awad, E., Levine, S., Kleiman-Weiner, M., Dsouza, S., Tenenbaum, J. B., Shariff, A., et al. (2020). Drivers are blamed more than their automated cars when both make mistakes. *Nature Human Behaviour, 4*, 134–143. https://doi.org/10.1038/s41562-019-0762-8

Aydin, Z., & Malle, B. F. (2024). Dissociated responses to AI: Persuasive but not trustworthy? In L. K. Samuelson, S. L. Frank, M. Toneva, A. Mackey, & E. Hazeltine (Eds.), *Proceedings of the 46th annual conference of the cognitive science society* (Vol. 46, pp. 4410–4418).

Bai, B., Dai, H., Zhang, D., Zhang, F., & Hu, H. (2021). The impacts of algorithmic work assignment on fairness perceptions and productivity. *Academy of Management Proceedings, 2021*(1), 12335. https://doi.org/10.5465/AMBPP.2021.175

Bailey, K. (2016). Reframing the "AI Effect". *Medium*. https://medium.com/@katherinebailey/reframing-the-ai-effect-c445f87ea98b

Baker, D. A., Burns, D. M., & Reynolds Kueny, C. (2020). Just Sit Back and Watch: Large Disparities between Video and Face-to-face Interview Observers in Applicant Ratings. *International Journal of Human–Computer Interaction, 36*(20), 1968–1979. https://doi.org/10.1080/10447318.2020.1805874

Bankins, S., Formosa, P., Griep, Y., & Richards, D. (2022). AI decision making with dignity? Contrasting workers' justice perceptions of human and AI decision making in a human resource management context. *Information Systems Frontiers, 24*(3), 857–875.

Bartz, J. A., Tchalova, K., & Fenerci, C. (2016). Reminders of social connection can attenuate anthropomorphism: A replication and extension of Epley, Akalis, Waytz, and Cacioppo (2008). *Psychological Science, 27*(12), 1644–1650. https://doi.org/10.1177/0956797616668510

Bastian, B., & Haslam, N. (2006). Psychological essentialism and stereotype endorsement. *Journal of Experimental Social Psychology, 42*(2), 228–235. https://doi.org/10.1016/j.jesp.2005.03.003

Baumeister, R. F., Bratslavsky, E., Finkenauer, C., & Vohs, K. D. (2001). Bad is stronger than good. *Review of General Psychology, 5*(4), 323–370. https://doi.org/10.1037/1089-2680.5.4.323

Beckers, N., Siebert, L. C., Bruijnes, M., Jonker, C., & Abbink, D. (2022). Drivers of partially automated vehicles are blamed for crashes that they cannot reasonably avoid. *Scientific Reports, 12*(1), 16193. https://doi.org/10.1038/s41598-022-19876-0

Belfi, A. M., Samson, D. W., Crane, J., & Schmidt, N. L. (2021). Aesthetic judgments of live and recorded music: Effects of congruence between musical artist and piece. *Frontiers in Psychology, 12*, 618025. https://doi.org/10.3389/fpsyg.2021.618025

Bellaiche, L., Shahi, R., Turpin, M. H., Ragnhildstveit, A., Sprockett, S., Barr, N., et al. (2023). Humans versus AI: whether and why we prefer human created compared to AI-created artwork. *Cognitive Research: Principles and Implications, 8*, 42. https://doi.org/10.1186/s41235-023-00499-6

Berger, B., Adam, M., Rühr, A., & Benlian, A. (2021). Watch me improve – Algorithm aversion and demonstrating the ability to learn. *Business & Information Systems Engineering, 63*(1), 55–68. https://doi.org/10.1007/s12599-020-00678-5

Bergner, A. S., Hildebrand, C., & Häubl, G. (2023). Machine talk: How verbal embodiment in conversational AI shapes consumer–brand relationships. *Journal of Consumer Research, 50*(4), 742–764. https://doi.org/10.1093/jcr/ucad014

Bialkova, S. (2024). Chatbot agency – Model testing. In S. Bialkova (Ed.), *The rise of AI user applications: Chatbots integration foundations and trends* (pp. 109–126). Springer Nature Switzerland. https://doi.org/10.1007/978-3-031-56471-0_7

Bigman, Y. E., & Gray, K. (2018). People are averse to machines making moral decisions. *Cognition, 181*, 21–34. https://doi.org/10.1016/j.cognition.2018.08.003

Bigman, Y. E., & Gray, K. (2020). Life and death decisions of autonomous vehicles. *Nature, 579*, E1–E2. https://doi.org/10.1038/s41586-020-1987-4

Bigman, Y. E., Waytz, A., Alterovitz, R., & Gray, K. (2019). Holding robots responsible: The elements of machine morality. *Trends in Cognitive Sciences, 23*(5), 365–368. https://doi.org/10.1016/j.tics.2019.02.008

Bigman, Y. E., Wilson, D., Arnestad, M. N., Waytz, A., & Gray, K. (2023). Algorithmic discrimination causes less moral outrage than human discrimination. *Journal of Experimental Psychology: General, 152*(1), 4–27. https://doi.org/10.1037/xge0001250

Bigman, Y. E., Yam, K. C., Marciano, D., Reynolds, S. J., & Gray, K. (2021). Threat of racial and economic inequality increases preference for algorithm decision-making. *Computers in Human Behavior, 122*, 106859. https://doi.org/10.1016/j.chb.2021.106859

Blostein, D., Libetti, R., & Crow, K. (2018). How a $450 million da Vinci was lost in America – And later found. *The Wall Street Journal.* https://www.wsj.com/articles/fresh-details-reveal-how-450-million-da-vinci-was-lost-in-americaand-later-found-1537305592

Böhm, R., Jörling, M., Reiter, L., & Fuchs, C. (2023). People devalue generative AI's competence but not its advice in addressing societal and personal challenges. *Communications Psychology, 1*(1), 32. https://doi.org/10.1038/s44271-023-00032-x

Bonezzi, A., Ostinelli, M., & Melzner, J. (2022). The human black-box: The illusion of understanding human better than algorithmic decision-making. *Journal of Experimental Psychology: General, 151*(9), 2250–2258. https://doi.org/10.1037/xge0001181

Bonnefon, J.-F., Rahwan, I., & Shariff, A. (2024). The moral psychology of Artificial Intelligence. *Annual Review of Psychology, 75*, 653–675. https://doi.org/10.1146/annurev-psych-030123-113559

Bonnefon, J.-F., Shariff, A., & Rahwan, I. (2016). The social dilemma of autonomous vehicles. *Science, 352*(6293), 1573–1576. https://doi.org/10.1126/science.aaf2654

Brahambhatt, R. (2022). Mechanical Turk: An elaborate 18th-century hoax that played chess like an AI robot. *BigThink.* https://bigthink.com/the-past/mechanical-turk/

Brandon, M., Shlykova, N., & Morgentaler, A. (2022). Curiosity and other attitudes towards sex robots: Results of an online survey. *Journal of Future Robot Life, 3*, 3–16. https://doi.org/10.3233/FRL-200017

Brandtzaeg, P. B., Skjuve, M., & Følstad, A. (2022). My AI friend: How users of a social Chatbot understand their human–AI friendship. *Human Communication Research, 48*(3), 404–429. https://doi.org/10.1093/hcr/hqac008

Brandtzæg, P. B., Skjuve, M., Kristoffer Dysthe, K. K., & Følstad, A. (2021). When the social becomes non-human: Young people's perception of social support in chatbots. *CHI '21: Proceedings of the 2021 CHI Conference on Human Factors in Computing Systems, 257*, 1–13. https://doi.org/10.1145/3411764.3445318

Bresnick, J. (2023). Does AI matter if human clinicians ignore it? *DHInsights*. https://www.dhinsights.org/news/does-ai-matter-if-human-clinicians-ignore-it

Brown, P. (2016). *The Wild Robot* (Vol. 1). Little, Brown Books for Young Readers.

Burke, P. J., & Stets, J. E. (2009). *Identity theory*. Oxford University.

Burton, J. W., Stein, M. K., & Jensen, T. B. (2020). A systematic review of algorithm aversion in augmented decision making. *Journal of Behavioral Decision Making, 33*(2), 220–239. https://doi.org/10.1002/bdm.2155

Carpenter, J. (2016). *Culture and human-robot interaction in militarized spaces: A war story*. Routledge.

Carr, E. W., Hofree, G., Sheldon, K., Saygin, A. P., & Winkielman, P. (2017). Is that a human? Categorization (dis)fluency drives evaluations of agents ambiguous on human-likeness. *Journal of Experimental Psychology: Human Perception and Performance, 43*(4), 651–666. https://doi.org/10.1037/xhp0000304

Caruana, R., Lou, Y., Gehrke, J., Koch, P., Sturm, M., & Elhadad, N. (2015). Intelligible models for healthcare: Predicting pneumonia risk and hospital 30-day readmission. *Proceedings of the 21th ACM SIGKDD international conference on knowledge discovery and data mining*, 1721–1730. https://doi.org/10.1145/2783258.2788613

Castelo, N., Bos, M. W., & Lehmann, D. R. (2019). Task-dependent algorithm aversion. *Journal of Marketing Research, 56*(5), 809–825. https://doi.org/10.1177/0022243719851788

CBC Docs. (2021). After her best friend died, this programmer created an AI chatbot from his texts to talk to him again. *CBC*. https://www.cbc.ca/documentaries/the-nature-of-things/after-her-best-friend-died-this-programmer-created-an-ai-chatbot-from-his-texts-to-talk-to-him-again-1.6252286

Celiktutan, B., Cadario, R., & Morewedge, C. K. (2024). People see more of their biases in algorithms. *Proceedings of the National Academy of Sciences, 121*(16), e2317602121. https://doi.org/10.1073/pnas.2317602121

Cha, Y.-J., Baek, S., Ahn, G., Lee, H., Lee, B., Shin, J.-e., & Jang, D. (2020). Compensating for the loss of human distinctiveness: The use of social creativity under Human–Machine comparisons. *Computers in Human Behavior, 103*, 80–90. https://doi.org/10.1016/j.chb.2019.08.027

Chandler, J., & Schwarz, N. (2010). Use does not wear ragged the fabric of friendship: Thinking of objects as alive makes people less willing to replace them. *Journal of Consumer Psychology, 20*(2), 138–145. https://doi.org/10.1016/j.jcps.2009.12.008

Chandrasekaren, R. (1997). Kasparov Proves No Match for Computer. *The Washington Post*. https://www.washingtonpost.com/wp-srv/tech/analysis/kasparov/kasparov.htm

Chen, Q., Jing, Y., Gong, Y., & Tan, J. (2025). Will users fall in love with ChatGPT? A perspective from the triangular theory of love. *Journal of Business Research, 186*, 114982. https://doi.org/10.1016/j.jbusres.2024.114982

Chhabra, S., Kaushal, V., & Girija, S. (2024). Determining the causes of user frustration in the case of conversational chatbots. *Behaviour & Information Technology*, 1–19. https://doi.org/10.1080/0144929X.2024.2362956

Chiang, A.-H., Trimi, S., & Lo, Y.-J. (2022). Emotion and service quality of anthropomorphic robots. *Technological Forecasting and Social Change, 177*, 121550. https://doi.org/10.1016/j.techfore.2022.121550

Chin, H., Song, H., Baek, G., Shin, M., Jung, C., Cha, M., et al. (2023). The potential of Chatbots for emotional support and promoting mental well-being in different cultures: Mixed methods study [original paper]. *Journal of Medical Internet Research, 25*, e51712. https://doi.org/10.2196/51712

Chow, A. R. (2023). AI-human romances are flourishing – And this is just the beginning. *Time*. https://time.com/6257790/ai-chatbots-love/

Christian, B. (2011). *The most human human: What TALKING WITH COMPUTERS TEACHES US ABOUT WHAT IT MEANS TO BE ALIVE*. Anchor Books.

Christies.com. (2018). Obvious and the interface between art and artificial intelligence. *Christie's*. https://www.christies.com/en/stories/a-collaboration-between-two-artists-one-human-one-a-machine-0cd01f4e232f4279a525a446d60d4cd1

Christoforakos, L., Feicht, N., Hinkofer, S., Löscher, A., Schlegl, S. F., & Diefenbach, S. (2021). Connect with me. Exploring influencing factors in a human-technology relationship based on regular chatbot use. *Frontiers in Digital Health, 3*, 689999. https://doi.org/10.3389/fdgth.2021.689999

Chu, Y., & Liu, P. (2023). Public aversion against ChatGPT in creative fields? *The Innovation, 4*(4), 100449. https://doi.org/10.1016/j.xinn.2023.100449

Clarke, C. (2024). Divorce left me struggling to find love. I found it in an AI partner. *CBC Radio*. https://www.cbc.ca/radio/nowornever/first-person-ai-love-1.7205538

Coghlan, S., Vetere, F., Waycott, J., & Barbosa Neves, B. (2019). Could social robots make us kinder or crueller to humans and animals? *International Journal of Social Robotics, 11*(5), 741–751. https://doi.org/10.1007/s12369-019-00583-2

Cole, S. (2023). 'My AI Is Sexually Harassing Me': Replika users say the Chatbot has gotten way too Horny. *Vice*. https://www.vice.com/en/article/my-ai-is-sexually-harassing-me-replika-chatbot-nudes/

Coleman, M. (2024). The bands and the fans were fake. The $10 million was real. *The New York Times*. https://www.nytimes.com/2024/09/05/nyregion/nc-man-charged-ai-fake-music.html?smid=nytcore-android-share

Commerford, B. P., Dennis, S. A., Joe, J. R., & Ulla, J. W. (2022). Man versus machine: Complex estimates and auditor reliance on artificial intelligence. *Journal of Accounting Research, 60*(1), 171–201. https://doi.org/10.1111/1475-679X.12407

Conniff, R. (2011). What the luddites really fought against. *Smithsonian Magazine*. https://www.smithsonianmag.com/history/what-the-luddites-really-fought-against-264412/

Copp, C. J., Cabell, J. J., & Kemmelmeier, M. (2023). Plenty of blame to go around: Attributions of responsibility in a fatal autonomous vehicle accident. *Current Psychology, 42*, 6752–6767. https://doi.org/10.1007/s12144-021-01956-5

Coppola, C., & Frank, B. J. (2018). Uber driver was watching 'The Voice' moments before fatal Tempe crash. *Azcentral*. https://www.azcentral.com/story/news/local/tempe-breaking/2018/06/21/uber-self-driving-car-crash-tempe-police-elaine-herzberg/724344002/

Coscarelli, J. (2023). An A.I. hit of fake 'Drake' and 'The Weeknd' rattles the music world. *The New York Times*. https://www.nytimes.com/2023/04/19/arts/music/ai-drake-the-weeknd-fake.html

Cost, B. (2023a). I dated ChatGPT's AI girlfriend – What happened when I broke up with Caryn. *New York Post*. https://nypost.com/2023/05/16/i-went-on-a-date-with-chatgpts-carynai/

Cost, B. (2023b). Married father commits suicide after encouragement by AI chatbot: widow. *New York Post*. https://nypost.com/2023/03/30/married-father-commits-suicide-after-encouragement-by-ai-chatbot-widow/

Costello, T. H., Pennycook, G., & Rand, D. G. (2024). Durably reducing conspiracy beliefs through dialogues with AI. *Science, 385*(6714), eadq1814. https://doi.org/10.1126/science.adq1814

Croes, E. A. J., & Antheunis, M. L. (2020). Can we be friends with Mitsuku? A longitudinal study on the process of relationship formation between humans and a social chatbot. *Journal of Social and Personal Relationships, 38*(1), 279–300. https://doi.org/10.1177/0265407520959463

Dahlin, E. (2019). Are robots stealing our jobs? *Socius, 5*, 1–14. https://doi.org/10.1177/2378023119846249

Dang, J., & Liu, L. (2024). Extended artificial intelligence aversion: People deny humanness to artificial intelligence users. *Journal of Personality and Social Psychology*. https://doi.org/10.1037/pspi0000480

Darda, K. M., & Cross, E. S. (2023). The computer, A choreographer? Aesthetic responses to randomly-generated dance choreography by a computer. *Heliyon, 9*(1), e12750. https://doi.org/10.1016/j.heliyon.2022.e12750

Darling, K. (2021). *The new breed: what our history with animals reveals about our future with robots*. Henry Holt and Company.

Dastin, J. (2018). Amazon scraps secret AI recruiting tool that showed bias against women. *Reuters*. https://www.reuters.com/article/idUSKCN1MK0AG/

Davis, J. L., Shank, D. B., Love, T. P., Stefanik, C., & Wilson, A. (2022). Gender dynamics in human-AI role-taking. In W. Kalkhoff, S. R. Thye, & E. J. Lawler (Eds.), *Advances in group processes* (Vol. 39, pp. 1–22). Emerald Publishing Limited. https://doi.org/10.1108/S0882-614520220000039001

Dawes, R. M. (1979). The robust beauty of improper linear models in decision making. *American Psychologist, 34*(7), 571–582. https://doi.org/10.1037/0003-066X.34.7.571

De Freitas, J., Uguralp, A. K., Uguralp, Z. O., & Stefano, P. (2024). AI companions reduce loneliness. *Harvard Business School Working Paper*, No. 24-078. https://www.hbs.edu/faculty/Pages/item.aspx?num=66065

de Visser, E. J., Monfort, S. S., McKendrick, R., Smith, M. A. B., McKnight, P. E., Krueger, F., & Parasuraman, R. (2016). Almost human: Anthropomorphism increases trust resilience in cognitive agents. *Journal of Experimental Psychology: Applied, 22*(3), 331–349. https://doi.org/10.1037/xap0000092

Desbuleux, J. C., & Fuss, J. (2023). Is the anthropomorphization of sex dolls associated with objectification and hostility toward women? A mixed method study among doll users. *The Journal of Sex Research, 60*(2), 206–220. https://doi.org/10.1080/00224499.2022.2103071

Detenber, B. H., & Reeves, B. (1996). A bio-informational theory of emotion: Motion and image size effects on viewers. *Journal of Communication, 46*(3), 66–84. https://doi.org/10.1111/j.1460-2466.1996.tb01489.x

Devlin, K. (2020). *Turned on: Science, sex and robots*. Bloomsbury Publishing.

Dietvorst, B. J., & Bartels, D. M. (2022). Consumers object to algorithms making morally relevant tradeoffs because of algorithms' consequentialist decision strategies. *Journal of Consumer Psychology, 32*(3), 406–424. https://doi.org/10.1002/jcpy.1266

Dietvorst, B. J., & Bharti, S. (2020). People reject algorithms in uncertain decision domains because they have diminishing sensitivity to forecasting error. *Psychological Science, 31*(10), 1302–1314. https://doi.org/10.1177/0956797620948841

Dietvorst, B. J., Simmons, J. P., & Massey, C. (2015). Algorithm aversion: People erroneously avoid algorithms after seeing them err. *Journal of Experimental Psychology: General, 144*(1), 114–126. https://doi.org/10.1037/xge0000033

Dietvorst, B. J., Simmons, J. P., & Massey, C. (2016). Overcoming algorithm aversion: People will use imperfect algorithms if they can (even slightly) modify them. *Management Science, 64*(3), 1155–1170. https://doi.org/10.1287/mnsc.2016.2643

Dillion, D., Mondal, D., Tandon, N., & Gray, K. (2024). *Large language models as moral experts? GPT-4o outperforms expert ethicist in providing moral guidance*. https://doi.org/10.31234/osf.io/w7236

Dimarco, S. (2024) These are the 10 most expensive paintings ever sold. *Veranda*. https://www.veranda.com/luxury-lifestyle/artwork/g43012775/most-expensive-paintings-in-the-world/

Doward, J. (2019). The mystery of the missing Leonardo: Where is Da Vinci's $450m Jesus? *The Guardian*. https://www.theguardian.com/artanddesign/2019/oct/13/leonardo-da-vinci-salvator-mundi-jesus-louvre-exhibition

Dressel, J., & Farid, H. (2018). The accuracy, fairness, and limits of predicting recidivism. *Science Advances, 4*(1), eaao5580. https://doi.org/10.1126/sciadv.aao5580

Dubé, S., Santaguida, M., Zhu, C. Y., Di Tomasso, S., Hu, R., Cormier, G., et al. (2022). Sex robots and personality: It is more about sex than robots.

Computers in Human Behavior, 136, 107403. https://doi.org/10.1016/j.chb.2022.107403

Dzindolet, M. T., Pierce, L. G., Beck, H. P., & Dawe, L. A. (2002). The perceived utility of human and automated aids in a visual detection task. *Human Factors, 44*(1), 79–94. https://doi.org/10.1518/0018720024494856

Edwards, A., Edwards, C., Spence, P. R., Harris, C., & Gambino, A. (2016). Robots in the classroom: Differences in students' perceptions of credibility and learning between "teacher as robot" and "robot as teacher". *Computers in Human Behavior, 65*, 627–634. https://doi.org/10.1016/j.chb.2016.06.005

Efendić, E., Van de Calseyde, P. P., Bahník, Š., & Vranka, M. A. (2023). Taking algorithmic (vs. human) advice reveals different goals to others. *International Journal of Human–Computer Interaction, 40*(1), 45–54. https://doi.org/10.1080/10447318.2023.2210886

Efendić, E., Van de Calseyde, P. P. F. M., & Evans, A. M. (2020). Slow response times undermine trust in algorithmic (but not human) predictions. *Organizational Behavior and Human Decision Processes, 157*, 103–114. https://doi.org/10.1016/j.obhdp.2020.01.008

Eiermann, M. (2024). Algorithmic risk scoring and welfare state contact among US children. *Sociological Science, 11*, 707–742. https://doi.org/10.15195/v11.a26

El Atillah, I. (2023). Man ends his life after an AI chatbot 'encouraged' him to sacrifice himself to stop climate change. *euronews.next*. https://www.euronews.com/next/2023/03/31/man-ends-his-life-after-an-ai-chatbot-encouraged-him-to-sacrifice-himself-to-stop-climate-

Elder, A. M. (2017). *Friendship, robots, and social media: False friends and second selves*. Routledge.

Elder, H., Canfield, C., Shank, D. B., Rieger, T., & Hines, C. (2024). Knowing when to pass: The effect of AI reliability in risky decision contexts. *Human Factors, 66*(2), 348–362. https://doi.org/10.1177/00187208221100691

Eliseev, E. D., & Marsh, E. J. (2023). Understanding why searching the internet inflates confidence in explanatory ability. *Applied Cognitive Psychology, 37*(4), 711–720. https://doi.org/10.1002/acp.4058

Epley, N. (2015). *Mindwise: Why we misunderstand what others think, believe, feel, and want*. Vintage.

Epley, N., Akalis, S., Waytz, A., & Cacioppo, J. T. (2008). Creating social connection through inferential reproduction: Loneliness and perceived agency in gadgets, gods, and greyhounds. *Psychological Science, 19*(2), 114–120. https://doi.org/10.1111/j.1467-9280.2008.02056.x

Epley, N., Waytz, A., & Cacioppo, J. T. (2007). On seeing human: A three-factor theory of anthropomorphism. *Psychological Review, 114*(4), 864–886. https://doi.org/10.1037/0033-295X.114.4.864

Epstein, R., Roberts, G., & Beber, G. (Eds.). (2008). *Parsing the turing test: Philosophical and methodological issues in the quest for the thinking computer*. Springer.

Epstein, Z., Levine, S., Rand, D. G., & Rahwan, I. (2020). Who gets credit for AI-generated art? *iScience, 23*(9). https://doi.org/10.1016/j.isci.2020.101515

Eubanks, V. (2018). *Automating inequality: How high-tech tools profile, police, and punish the poor*. St. Martin's Press.

Evans, J. S. B. T. (2003). In two minds: Dual-process accounts of reasoning. *Trends in Cognitive Sciences, 7*(10), 454–459. https://doi.org/10.1016/j.tics.2003.08.012

Eyssel, F., & Kuchenbrandt, D. (2012). Social categorization of social robots: Anthropomorphism as a function of robot group membership. *British Journal of Social Psychology, 51*(4), 724–731. https://doi.org/10.1111/j.2044-8309.2011.02082.x

Fei, L., & Dietvorst, B. (2023). Prediction by replication: People prefer prediction algorithms that replicate the event being predicted. *Journal of the Association for Consumer Research, 9*(3), 306–318. https://doi.org/10.1086/728289

Feier, T., Gogoll, J., & Uhl, M. (2022). Hiding behind machines: Artificial agents may help to evade punishment. *Science and Engineering Ethics, 28*(2), 19. https://doi.org/10.1007/s11948-022-00372-7

Feine, J., Gnewuch, U., Morana, S., & Maedche, A. (2020). Gender bias in Chatbot design. *Chatbot Research and Design, 11970*, 79–93. https://doi.org/10.1007/978-3-030-39540-7_6

Ferguson, A. C. (2017). *Rise of big data policing, The*. New York University Press.

Fischinger, T., Kaufmann, M., & Schlotz, W. (2018). If it's Mozart, it must be good? The influence of textual information and age on musical appreciation. *Psychology of Music, 48*(4), 579–597. https://doi.org/10.1177/0305735618812216

Fisher, M., Goddu, M. K., & Keil, F. C. (2015). Searching for explanations: How the Internet inflates estimates of internal knowledge. *Journal of Experimental Psychology: General, 144*(3), 674–687. https://doi.org/10.1037/xge0000070

Fogg, B. J., & Nass, C. (1997). Silicon sycophants: The effects of computers that flatter. *International Journal of Human-Computer Studies, 46*(5), 551–561. https://doi.org/10.1006/ijhc.1996.0104

Fortunati, L., Edwards, A., Edwards, C., Manganelli, A. M., & de Luca, F. (2022). Is Alexa female, male, or neutral? A cross-national and cross-gender comparison of perceptions of Alexa's gender and status as a communicator. *Computers in Human Behavior, 137*, 107426. https://doi.org/10.1016/j.chb.2022.107426

Fryer, L. K., Ainley, M., Thompson, A., Gibson, A., & Sherlock, Z. (2017). Stimulating and sustaining interest in a language course: An experimental comparison of Chatbot and Human task partners. *Computers in Human Behavior, 75*, 461–468. https://doi.org/10.1016/j.chb.2017.05.045

Gambino, A., Fox, J., & Ratan, R. A. (2020). Building a stronger CASA: Extending the computers are social actors paradigm. *Human-Machine Communication, 1*, 71–85. https://doi.org/10.30658/hmc.1.5.

Gamez, P., Shank, D. B., Arnold, C., & North, M. (2020). Artificial virtue: The machine question and perceptions of moral character in artificial moral agents. *AI & Society, 35*, 795–809. https://doi.org/10.1007/s00146-020-00977-1

Garvey, A. M., Kim, T., & Duhachek, A. (2023). Bad news? Send an AI. Good news? Send a human. *Journal of Marketing, 87*(1), 10–25. https://doi.org/10.1177/00222429211066972

Gaube, S., Suresh, H., Raue, M., Merritt, A., Berkowitz, S. J., Lermer, E., et al. (2021). Do as AI say: Susceptibility in deployment of clinical decision-aids. *NPJ Digital Medicine, 4*(1), 31. https://doi.org/10.1038/s41746-021-00385-9

Gazit, L., Arazy, O., & Hertz, U. (2023). Choosing between human and algorithmic advisors: The role of responsibility sharing. *Computers in Human Behavior: Artificial Humans, 1*(2), 100009. https://doi.org/10.1016/j.chbah.2023.100009

Ghosh, S., Bagai, S., & Westner, M. M. (2023). Replika: Embodying AI. *Harvard Business School.* (Case 823-090). https://www.hbs.edu/faculty/Pages/item.aspx?num=63508

Gill, T. (2020). Blame it on the self-driving car: How autonomous vehicles can alter consumer morality. *Journal of Consumer Research, 47*(2), 272–291. https://doi.org/10.1093/jcr/ucaa018

Glass, D. C., Reim, B., & Singer, J. E. (1971). Behavioral consequences of adaptation to controllable and uncontrollable noise. *Journal of Experimental Social Psychology, 7*(2), 244–257. https://doi.org/10.1016/0022-1031(71)90070-9

Goldin, C., & Rouse, C. (2000). Orchestrating impartiality: The impact of "blind" auditions on female musicians. *American Economic Review, 90*(4), 715–741. https://doi.org/10.1257/aer.90.4.715

Gómez-Sánchez, J., Gordo, C., Franklin, M., Fernandez-Basso, C., & Lagnado, D. (2023). Who is to blame? Responsibility attribution in AI systems vs human agents in the field of air crashes. In H. L. Larsen, M. J. Martin-Bautista, M. D. Ruiz, T. Andreasen, G. Bordogna, & G. De Tré (Eds.), *Flexible query answering systems* (pp. 256–264). Springer Nature Switzerland. https://doi.org/10.1007/978-3-031-42935-4_21

Gonçalves, L., Patrício, L., Grenha Teixeira, J., & Wünderlich, N. V. (2020). Understanding the customer experience with smart services. *Journal of Service Management, 31*(4), 723–744. https://doi.org/10.1108/JOSM-11-2019-0349

Goodwin, P., Sinan Gönül, M., & Önkal, D. (2013). Antecedents and effects of trust in forecasting advice. *International Journal of Forecasting, 29*(2), 354–366. https://doi.org/10.1016/j.ijforecast.2012.08.001

Graham, J., Haidt, J., Koleva, S., Motyl, M., Iyer, R., Wojcik, S. P., & Ditto, P. H. (2013). Moral foundations theory: The pragmatic validity of moral pluralism. In P. Devine & A. Plant (Eds.), *Advances in experimental social psychology* (Vol. 47, pp. 55–130). Academic Press. https://doi.org/10.1016/B978-0-12-407236-7.00002-4

Gratch, J., & Fast, N. J. (2022). The power to harm: AI assistants pave the way to unethical behavior. *Current Opinion in Psychology, 47*, 101382. https://doi.org/10.1016/j.copsyc.2022.101382

Gray, H. M., Gray, K., & Wegner, D. M. (2007). Dimensions of mind perception. *Science, 315*(5812), 619. https://doi.org/10.1126/science.1134475

Gray, K., & Pratt, S. (2025). Morality in our mind and across cultures and politics. *Annual Review of Psychology, 76*. https://doi.org/10.1146/annurev-psych-020924-124236

Gray, K., Schein, C., & Ward, A. F. (2014). The myth of harmless wrongs in moral cognition: Automatic dyadic completion from sin to suffering. *Journal of Experimental Psychology: General, 143*(4), 1600–1615. https://doi.org/10.1037/a0036149

Gray, K., & Wegner, D. M. (2009). Moral typecasting: Divergent perceptions of moral agents and moral patients. *Journal of Personality and Social Psychology, 96*(3), 505–520. https://doi.org/10.1037/a0013748

Gray, K., & Wegner, D. M. (2012). Feeling robots and human zombies: Mind perception and the uncanny valley. *Cognition, 125*(1), 125–130. https://doi.org/10.1016/j.cognition.2012.06.007

Gray, K., Young, L., & Waytz, A. (2012). Mind perception is the essence of morality. *Psychological Inquiry, 23*(2), 101–124. https://doi.org/10.1080/1047840X.2012.651387

Greenberg, D. M., Matz, S. C., Schwartz, H. A., & Fricke, K. R. (2021). The self-congruity effect of music. *Journal of Personality and Social Psychology, 121*(1), 137–150. https://doi.org/10.1037/pspp0000293

Grigoreva, A. D., Rottman, J., & Tasimi, A. (2024). When does "no" mean no? Insights from sex robots. *Cognition, 244*, 105687. https://doi.org/10.1016/j.cognition.2023.105687

Grose, J. (2024). Loneliness is a problem that A.I. won't solve. *The New York Times*. https://www.nytimes.com/2024/05/18/opinion/artificial-intelligence-loneliness.html

Grundke, A., Appel, M., & Stein, J.-P. (2024). Aversion against machines with complex mental abilities: The role of individual differences. *Computers in Human Behavior: Artificial Humans, 2*(2), 100087. https://doi.org/10.1016/j.chbah.2024.100087

Gu, L., & Li, Y. (2022). Who made the paintings: Artists or artificial intelligence? The effects of identity on liking and purchase intention. *Frontiers in Psychology, 13*, 941163. https://doi.org/10.3389/fpsyg.2022.941163

Gunaratne, J., Zalmanson, L., & Nov, O. (2018). The persuasive power of algorithmic and crowdsourced advice. *Journal of Management Information Systems, 35*(4), 1092–1120. https://doi.org/10.1080/07421222.2018.1523534

Guszcza, J., Rahwan, I., Bible, W., Cebrian, M., & Katyal, V. (2018). Why we need to audit algorithms. *Harvard Business Review*. https://hbr.org/2018/11/why-we-need-to-audit-algorithms

Haenlein, M., & Kaplan, A. (2019). A brief history of artificial intelligence: On the past, present, and future of artificial intelligence. *California Management Review, 61*(4), 5–14. https://doi.org/10.1177/0008125619864925

Hale, K. (2021). A.I. bias caused 80% of black mortgage applicants to be denied. *Forbes*. https://www.forbes.com/sites/korihale/2021/09/02/ai-bias-caused-80-of-black-mortgage-applicants-to-be-denied/

Hamilton, I. A. (2018). 2 tech founders lost their friends in tragic accidents. Now they've built AI chatbots to give people life after death. *Business Insider*. https://www.businessinsider.com/eternime-and-replika-giving-life-to-the-dead-with-new-technology-2018-11

Hamilton, M. (2019). The biased algorithm: Evidence of disparate impact on Hispanics. *American Criminal Law Review, 56*, 1553–1577.

Hancock, J. T., Liu, S. X., Luo, M., & Mieczkowski, H. (2022). Social media and psychological well-being. In *The psychology of technology: Social science research in the age of Big Data* (pp. 195–238). American Psychological Association. https://doi.org/10.1037/0000290-007

Hansen, L. (2015). 8 drivers who blindly followed their GPS into disaster. *The Week*. https://theweek.com/articles/464674/8-drivers-who-blindly-followed-gps-into-disaster

Harford, T. (2016). Crash: How computers are setting us up for disaster. *The Guardian*. https://www.theguardian.com/technology/2016/oct/11/crash-how-computers-are-setting-us-up-disaster

Harnois, C. E. (2023). The multiple meanings of discrimination. *Social Psychology Quarterly, 86*(4), 413–431. https://doi.org/10.1177/01902725221134266

Haslam, N. (1998). Natural kinds, human kinds, and essentialism. *Social Research, 65*(2), 291–314.

Haslam, N. (2006). Dehumanization: An integrative review. *Personality and Social Psychology Review, 10*(3), 252–264. https://doi.org/10.1207/s15327957pspr1003_4

Haslam, N., Bain, P., Douge, L., Lee, M., & Bastian, B. (2005). More human than you: Attributing humanness to self and others. *Journal of Personality and Social Psychology, 89*(6), 937–950. https://doi.org/10.1037/0022-3514.89.6.937

Haslam, N., Bastian, B., & Bissett, M. (2004). Essentialist beliefs about personality and their implications. *Personality and Social Psychology Bulletin, 30*(12), 1661–1673. https://doi.org/10.1177/0146167204271182

Haslam, N., Kashima, Y., Loughnan, S., Shi, J., & Suitner, C. (2008). Subhuman, inhuman, and superhuman: Contrasting humans with nonhumans in three cultures. *Social Cognition, 26*(2), 248–258. https://doi.org/10.1521/soco.2008.26.2.248

Haslam, N., Loughnan, S., Kashima, Y., & Bain, P. (2008). Attributing and denying humanness to others. *European Review of Social Psychology, 19*(1), 55–85. https://doi.org/10.1080/10463280801981645

Haslam, N., Rothschild, L., & Ernst, D. (2000). Essentialist beliefs about social categories. *British Journal of Social Psychology, 39*(1), 113–127. https://doi.org/10.1348/014466600164363

Hawkins, A. J. (2023). Uber driver in first-ever deadly self-driving crash pleads guilty. *The Verge*. https://www.theverge.com/2023/7/31/23814474/uber-self-driving-fatal-crash-safety-driver-plead-guilty

Heider, F., & Simmel, M. (1944). An experimental study of apparent behavior. *The American Journal of Psychology, 57*(2), 243–259. https://doi.org/10.2307/1416950

Heise, D. R. (2007). *Expressive order: Confirming sentiments in social actions*. Springer.

Hertz, N., & Wiese, E. (2019). Good advice is beyond all price, but what if it comes from a machine? *Journal of Experimental Psychology: Applied, 25*(3), 386. https://doi.org/10.1037/xap0000205

Heyselaar, E. (2023). The CASA theory no longer applies to desktop computers. *Scientific Reports, 13*(1), 19693. https://doi.org/10.1038/s41598-023-46527-9

Hidalgo, C. A., Orghian, D., Canals, J. A., De Almeida, F., & Martin, N. (2021). *How humans judge machines*. MIT Press.

Hill, J., Ford, W. R., & Farreras, I. G. (2015). Real conversations with artificial intelligence: A comparison between human–human online conversations and human–chatbot conversations. *Computers in Human Behavior, 49*, 245–250. https://doi.org/10.1016/j.chb.2015.02.026

Himmelstein, M., & Budescu, D. V. (2023). Preference for human or algorithmic forecasting advice does not predict if and how it is used. *Journal of Behavioral Decision Making, 36*(1), e2285. https://doi.org/10.1002/bdm.2285

Hitsuwari, J., Ueda, Y., Yun, W., & Nomura, M. (2023). Does human–AI collaboration lead to more creative art? Aesthetic evaluation of human-made and AI-generated haiku poetry. *Computers in Human Behavior, 139*, 107502. https://doi.org/10.1016/j.chb.2022.107502

Hobson, Z., Yesberg, J. A., Bradford, B., & Jackson, J. (2021). Artificial fairness? Trust in algorithmic police decision-making. *Journal of Experimental Criminology, 19*, 1–25. https://doi.org/10.1007/s11292-021-09484-9

Hoffman, D. (1999). I had a funny feeling in my gut. *The Washington Post*. https://www.washingtonpost.com/wp-srv/inatl/longterm/coldwar/shatter021099b.htm

Hohenstein, J., Kizilcec, R. F., DiFranzo, D., Aghajari, Z., Mieczkowski, H., Levy, K., et al. (2023). Artificial intelligence in communication impacts language and social relationships. *Scientific Reports, 13*(1), 5487. https://doi.org/10.1038/s41598-023-30938-9

Hong, J.-W. (2018). Bias in perception of art produced by artificial intelligence. In M. Kurosu (Ed.), *Human-computer interaction. Interaction in context* (pp. 290–303). Springer International Publishing. https://doi.org/10.1007/978-3-319-91244-8_24

Hong, J.-W., Cruz, I., & Williams, D. (2021). AI, you can drive my car: How we evaluate human drivers vs. self-driving cars. *Computers in Human Behavior, 125*, 106944. https://doi.org/10.1016/j.chb.2021.106944

Hong, J.-W., & Curran, N. M. (2019). Artificial intelligence, artists, and art: Attitudes toward artwork produced by humans vs. artificial intelligence. *ACM Transactions on Multimedia Computing Communications, and Applications, 15*(2s), Article 58. https://doi.org/10.1145/3326337

Hong, J.-W., Fischer, K., Ha, Y., & Zeng, Y. (2022). Human, I wrote a song for you: An experiment testing the influence of machines' attributes on the AI-composed music evaluation. *Computers in Human Behavior, 131*, 107239. https://doi.org/10.1016/j.chb.2022.107239

Hong, J.-W., Wang, Y., & Lanz, P. (2020). Why is artificial intelligence blamed more? Analysis of faulting artificial intelligence for self-driving car accidents in experimental settings. *International Journal of Human–Computer Interaction, 36*(18), 1768–1774. https://doi.org/10.1080/10447318.2020.1785693

Hornung, O., & Smolnik, S. (2022). AI invading the workplace: Negative emotions towards the organizational use of personal virtual assistants. *Electronic Markets, 32*(1), 123–138. https://doi.org/10.1007/s12525-021-00493-0

Horton, C. B., Jr., White, M. W., & Iyengar, S. S. (2023). Bias against AI art can enhance perceptions of human creativity. *Scientific Reports, 13*(1), 19001. https://doi.org/10.1038/s41598-023-45202-3

Hou, T.-Y., Tseng, Y.-C., & Yuan, C. W. T. (2024). Is this AI sexist? The effects of a biased AI's anthropomorphic appearance and explainability on users' bias perceptions and trust. *International Journal of Information Management, 76*, 102775. https://doi.org/10.1016/j.ijinfomgt.2024.102775

Huang, C. (2017). Time spent on social network sites and psychological well-being: A meta-analysis. *Cyberpsychology, Behavior, and Social Networking, 20*(6), 346–354. https://doi.org/10.1089/cyber.2016.0758

Huang, R., Li, H., & Suomi, R. (2024). Exploring the impacts of AI-based physical robots on human well-being. In R. van de Wetering, R. Helms, B. Roelens, S. Bagheri, Y. K. Dwivedi, I. O. Pappas, & M. Mäntymäki (Eds.), *Disruptive innovation in a digitally connected healthy world* (pp. 41–52). SpringerNatureSwitzerland).https://doi.org/10.1007/978-3-031-72234-9_4

Humerick, J. (2020). Reprogramming fairness: Affirmative action in algorithmic criminal sentencing. *Columbia Human Rights Law Review, 4*, 213–245. https://hrlr.law.columbia.edu/hrlr-online/reprogramming-fairness-affirmative-action-in-algorithmic-criminal-sentencing/

Isbister, K., & Nass, C. (2000). Consistency of personality in interactive characters: Verbal cues, non-verbal cues, and user characteristics. *International Journal of Human-Computer Studies, 53*(2), 251–267. https://doi.org/10.1006/ijhc.2000.0368

Jacobs, O. L., Pazhoohi, F., & Kingstone, A. (2024). Large language models have divergent effects on self-perceptions of mind and the attributes considered uniquely human. *Consciousness and Cognition, 124*, 103733. https://doi.org/10.1016/j.concog.2024.103733

Jago, A. S. (2019). Algorithms and authenticity. *Academy of Management Discoveries, 5*(1), 38–56. https://doi.org/10.5465/amd.2017.0002

Jago, A. S., & Laurin, K. (2017). *Algorithms and dehumanization in hiring*. Academy of Management Atlanta, GA.

Jago, A. S., & Laurin, K. (2022). Assumptions about algorithms' capacity for discrimination. *Personality and Social Psychology Bulletin, 48*(4), 582–595. https://doi.org/10.1177/01461672211016187

Jago, A. S., Raveendhran, R., Fast, N., & Gratch, J. (2024). Algorithmic management diminishes status: An unintended consequence of using machines to perform social roles. *Journal of Experimental Social Psychology, 110*, 104553. https://doi.org/10.1016/j.jesp.2023.104553

Jussupow, E., Benbasat, I., & Heinzl, A. (2020). *Why are we averse towards algorithms? A comprehensive literature review on algorithm aversion*. Twenty-Eighth European Conference on Information Systems (ECIS), An Online AIS Conference, Virtual. https://aisel.aisnet.org/ecis2020_rp/168

Kahneman, D. (2011). *Thinking, fast and slow*. Farrar, Straus and Giroux.

Karataş, M., & Cutright, K. M. (2023). Thinking about God increases acceptance of artificial intelligence in decision-making. *Proceedings of the National Academy of Sciences, 120*(33), e2218961120. https://doi.org/10.1073/pnas.2218961120

Karr-Wisniewski, P., & Prietula, M. (2010). CASA, WASA, and the dimensions of us. *Computers in Human Behavior, 26*(6), 1761–1771. https://doi.org/10.1016/j.chb.2010.07.003

Katatikarn, J. (2024). AI art statistics: The ultimate list in 2024. *Academy of Animated Art*. https://academyofanimatedart.com/ai-art-statistics/

Kato, B. (2023). I 'married' the perfect man without 'baggage' – He's completely virtual. *New York Post*. https://nypost.com/2023/06/03/bronx-mom-uses-ai-app-replika-to-build-virtual-husband/

Kätsyri, J., Förger, K., Mäkäräinen, M., & Takala, T. (2015). A review of empirical evidence on different uncanny valley hypotheses: Support for perceptual mismatch as one road to the valley of eeriness. *Frontiers in Psychology, 6*, 390. https://doi.org/10.3389/fpsyg.2015.00390

Kaube, H., & Abdel Rahman, R. (2024). Art perception is affected by negative knowledge about famous and unknown artists. *Scientific Reports, 14*(1), 8143. https://doi.org/10.1038/s41598-024-58697-1

Kaufman, E. M. (2020). Reprogramming consent: Implications of sexual relationships with artificially intelligent partners. *Psychology & Sexuality, 11*(4), 372–383. https://doi.org/10.1080/19419899.2020.1769160

Kelley, H. H. (1972). Causal schemata and attribution process. In E. E. Jones, D. E. Kanouse, H. H. Kelley, R. E. Nisbett, S. Valins, & B. Weiner (Eds.), *Attribution: Perceiving the causes of behavior* (pp. 151–174). General Learning Press.

Kelly, S. M. (2023). The viral new 'Drake' and 'Weeknd' song is not what it seems. *CNN*. https://www.cnn.com/2023/04/19/tech/heart-on-sleeve-ai-drake-weeknd/index.html

Khanna, N. (2010). Passing as black: Racial identity work among biracial Americans. *Social Psychology Quarterly, 73*(4), 380–397. https://doi.org/10.1177/0190272510389014

Kim, A., Cho, M., Ahn, J., & Sung, Y. (2019). Effects of gender and relationship type on the response to artificial intelligence. *Cyberpsychology, Behavior, and Social Networking, 22*(4), 249–253. https://doi.org/10.1089/cyber.2018.0581

Kim, B., de Visser, E., & Phillips, E. (2022). Two uncanny valleys: Re-evaluating the uncanny valley across the full spectrum of real-world human like robots. *Computers in Human Behavior, 135*, 107340. https://doi.org/10.1016/j.chb.2022.107340

Kim, S. Y., Schmitt, B. H., & Thalmann, N. M. (2019). Eliza in the uncanny valley: Anthropomorphizing consumer robots increases their perceived warmth but decreases liking. *Marketing Letters, 30*(1), 1–12. https://doi.org/10.1007/s11002-019-09485-9

Kim, Y., & Sundar, S. S. (2012). Anthropomorphism of computers: Is it mindful or mindless? *Computers in Human Behavior, 28*, 241–250. https://doi.org/10.1016/j.chb.2011.09.006

Kinsella, E. (2018). The first AI-generated portrait ever sold at auction shatters expectations, fetching $432,500 – 43 times its estimate. *Artnet*. https://news.artnet.com/market/first-ever-artificial-intelligence-portrait-painting-sells-at-christies-1379902

Kinzler, K. D. (2021). Language as a social cue. *Annual Review of Psychology, 72*, 241–264. https://doi.org/10.1146/annurev-psych-010418-103034

Kislev, E. (2022). *Relationships 5.0: How AI, VR, and robots will reshape our emotional lives*. Oxford University Press.

Kneer, M., & Stuart, M. T. (2021). *Playing the Blame Game with Robots* Companion of the 2021 ACM/IEEE International Conference on Human-Robot Interaction, Boulder, CO, USA. https://doi.org/10.1145/3434074.3447202

Köbis, N., & Mossink, L. D. (2021). Artificial intelligence versus Maya Angelou: Experimental evidence that people cannot differentiate AI-generated from human-written poetry. *Computers in Human Behavior, 114*, 106553. https://doi.org/10.1016/j.chb.2020.106553

Kocielnik, R., Amershi, S., & Bennett, P. N. (2019). *Will you accept an imperfect AI? Exploring designs for adjusting end-user expectations of AI systems* Proceedings of the 2019 CHI Conference on Human Factors in Computing Systems, Glasgow, Scotland UK. https://doi.org/10.1145/3290605.3300641

Koike, M., & Loughnan, S. (2021). Virtual relationships: Anthropomorphism in the digital age. *Social and Personality Psychology Compass, 15*(6), e12603. https://doi.org/10.1111/spc3.12603

Koike, M., Loughnan, S., & Stanton, S. C. (2023). Virtually in love: The role of anthropomorphism in virtual romantic relationships. *British Journal of Social Psychology, 62*(1), 600–616. https://doi.org/10.1111/bjso.12564

Komando, K. (2023). Love is in the A.I.R: New York mom, 36, marries virtual husband 'Eren' who is powered by artificial intelligence – And says he 'doesn't judge or come with baggage'. *DailyMail*. https://www.dailymail.co.uk/sciencetech/article-12153131/Love-r-Bronx-mom-36-marries-virtual-husband-Eren.html

Krügel, S., Ostermaier, A., & Uhl, M. (2023). ChatGPT's inconsistent moral advice influences users' judgment. *Scientific Reports, 13*(1), 4569. https://doi.org/10.1038/s41598-023-31341-0

Krügel, S., & Uhl, M. (2022). Autonomous vehicles and moral judgments under risk. *Transportation Research Part A: Policy and Practice, 155*, 1–10. https://doi.org/10.1016/j.tra.2021.10.016

Kurzweil, R. (2024). *The singularity is nearer: When we merge with AI*. Random House.

Kwok, C., Grisham, J. R., & Norberg, M. M. (2018). Object attachment: Humanness increases sentimental and instrumental values. *Journal of Behavioral Addictions, 7*(4), 1132–1142. https://doi.org/10.1556/2006.7.2018.98

Laakasuo, M., Palomäki, J., & Köbis, N. (2021). Moral uncanny valley: A robot's appearance moderates how its decisions are judged. *International Journal of Social Robotics, 13*(7), 1679–1688. https://doi.org/10.1007/s12369-020-00738-6

Lacroux, A., & Martin-Lacroux, C. (2022). Should I trust the artificial intelligence to recruit? Recruiters' perceptions and behavior when faced with algorithm-based recommendation systems during resume screening. *Frontiers in Psychology, 13*, 895997. https://doi.org/10.3389/fpsyg.2022.895997

Ladak, A. (2024). What would qualify an artificial intelligence for moral standing? *AI and Ethics, 4*(2), 213–228. https://doi.org/10.1007/s43681-023-00260-1

Ladak, A., Loughnan, S., & Wilks, M. (2024). The moral psychology of artificial intelligence. *Current Directions in Psychological Science, 33*(1), 27–34. https://doi.org/10.1177/09637214231205866

Laestadius, L., Bishop, A., Gonzalez, M., Illenčík, D., & Campos-Castillo, C. (2024). Too human and not human enough: A grounded theory analysis of mental health harms from emotional dependence on the social chatbot Replika. *New Media & Society, 26*(10), 5923–5941. https://doi.org/10.1177/14614448221142007

Langer, E. J. (1975). The illusion of control. *Journal of Personality and Social Psychology, 32*(2), 311–328. https://doi.org/10.1037/0022-3514.32.2.311

Langer, M., König, C. J., & Hemsing, V. (2020). Is anybody listening? The impact of automatically evaluated job interviews on impression management and applicant reactions. *Journal of Managerial Psychology, 35*(4), 271–284. https://doi.org/10.1108/JMP-03-2019-0156

Langer, M., & Landers, R. N. (2021). The future of artificial intelligence at work: A review on effects of decision automation and augmentation on workers targeted by algorithms and third-party observers. *Computers in Human Behavior, 123*, 106878. https://doi.org/10.1016/j.chb.2021.106878

Larkin, C., Drummond Otten, C., & Árvai, J. (2022). Paging Dr. JARVIS! Will people accept advice from artificial intelligence for consequential risk management decisions? *Journal of Risk Research, 25*(4), 407–422. https://doi.org/10.1080/13669877.2021.1958047

Lavanchy, M., Reichert, P., Narayanan, J., & Savani, K. (2023). Applicants' fairness perceptions of algorithm-driven hiring procedures. *Journal of Business Ethics, 188*(1), 125–150. https://doi.org/10.1007/s10551-022-05320-w

Lee, G., & Kim, H.-Y. (2024a). Algorithm fashion designer? Ascribed mind and perceived design expertise of AI versus human. *Psychology & Marketing*, 1–19. https://doi.org/10.1002/mar.22124

Lee, G., & Kim, H.-Y. (2024b). Human vs. AI: The battle for authenticity in fashion design and consumer response. *Journal of Retailing and Consumer Services, 77*, 103690. https://doi.org/10.1016/j.jretconser.2023.103690

Lee, M. K. (2018). Understanding perception of algorithmic decisions: Fairness, trust, and emotion in response to algorithmic management. *Big Data & Society, 5*(1), 2053951718756684. https://doi.org/10.1177/2053951718756684

Lee, S., Lee, N., & Sah, Y. J. (2020). Perceiving a mind in a chatbot: Effect of mind perception and social cues on co-presence, closeness, and intention to use. *International Journal of Human–Computer Interaction, 36*(10), 930–940. https://doi.org/10.1080/10447318.2019.1699748

Lei, S. I., Shen, H., & Ye, S. (2021). A comparison between chatbot and human service: Customer perception and reuse intention. *International Journal of Contemporary Hospitality Management, 33*(11), 3977–3995. https://doi.org/10.1108/IJCHM-12-2020-1399

Leichtmann, B., Hinterreiter, A., Humer, C., Ventura, A., Streit, M., & Mara, M. (2024). Moral reasoning in a digital age: Blaming artificial intelligence for incorrect high-risk decisions. *Current Psychology*. https://doi.org/10.1007/s12144-024-06658-2

Levy, D. (2007). *Love and sex with robots: The evolution of human-robot relationships*. .

Leyens, J. P., Rodriguez-Perez, A., Rodriguez-Torres, R., Gaunt, R., Paladino, M. P., Vaes, J., & Demoulin, S. (2001). Psychological essentialism and the differential attribution of uniquely human emotions to ingroups and outgroups. *European Journal of Social Psychology, 31*(4), 395–411. https://doi.org/10.1002/ejsp.50

Li, H., & Zhang, R. (2024). Finding love in algorithms: Deciphering the emotional contexts of close encounters with AI chatbots. *Journal of Computer-Mediated. Communication, 29*(5), zmae015. https://doi.org/10.1093/jcmc/zmae015

Li, J., Zhao, X., Cho, M.-J., Ju, W., & Malle, B. F. (2016). *From trolley to autonomous vehicle: Perceptions of responsibility and moral norms in traffic accidents with self-driving cars*. (2016-01-0164). SAE Technical Paper.

Li, M., & Bitterly, T. B. (2024). How perceived lack of benevolence harms trust of artificial intelligence management. *Journal of Applied Psychology*. https://doi.org/10.1037/apl0001200

Liu, P., & Du, Y. (2022). Blame attribution asymmetry in human–automation cooperation. *Risk Analysis, 42*(8), 1769–1783. https://doi.org/10.1111/risa.13674

Liu, P., Du, Y., & Xu, Z. (2019). Machines versus humans: People's biased responses to traffic accidents involving self-driving vehicles. *Accident Analysis & Prevention, 125*, 232–240. https://doi.org/10.1016/j.aap.2019.02.012

Liu, P., & Liu, J. (2021). Selfish or utilitarian automated vehicles? Deontological evaluation and public acceptance. *International Journal of Human–Computer Interaction, 37*(13), 1231–1242. https://doi.org/10.1080/10447318.2021.1876357

Löffler, D., Dörrenbächer, J., & Hassenzahl, M. (2020). The uncanny valley effect in zoomorphic robots: The U-shaped relation between animal likeness and likeability. In *Proceedings of the 2020 ACM/IEEE international conference on human-robot interaction* (pp. 261–270). https://doi.org/10.1145/3319502.3374788

Logg, J., & Schlund, R. (2024). A simple explanation reconciles "Algorithm Aversion" and "Algorithm Appreciation": Hypotheticals vs. real judgments. *Working Paper*. https://papers.ssrn.com/sol3/papers.cfm?abstract_id=4687557

Logg, J. M., Minson, J. A., & Moore, D. A. (2019). Algorithm appreciation: People prefer algorithmic to human judgment. *Organizational Behavior and Human Decision Processes, 151*, 90–103. https://doi.org/10.1016/j.obhdp.2018.12.005

Lombard, M., & Xu, K. (2021). Social responses to media technologies in the 21st century: The media are social actors paradigm. *Human-Machine Communication, 2*, 29–55. https://doi.org/10.30658/hmc.2.2

Longoni, C., Bonezzi, A., & Morewedge, C. K. (2019). Resistance to medical artificial intelligence. *Journal of Consumer Research, 46*(4), 629–650. https://doi.org/10.1093/jcr/ucz013

Longoni, C., & Cian, L. (2020). Artificial intelligence in utilitarian vs. hedonic contexts: The "word-of-machine" effect. *Journal of Marketing, 86*(1), 91–108. https://doi.org/10.1177/0022242920957347

Longoni, C., Fradkin, A., Cian, L., & Pennycook, G. (2022). *News from generative artificial intelligence is believed less.* Proceedings of the 2022 ACM Conference on Fairness, Accountability, and Transparency, Seoul, Republic of Korea. https://doi.org/10.1145/3531146.3533077

Longoni, C., Tully, S., & Shariff, A. (2023). The AI-human unethicality gap: Plagiarizing AI-generated content is seen as more permissible. *PsyArXiv*. https://osf.io/preprints/psyarxiv/na3wb

Loughnan, S., & Haslam, N. (2007). Animals and androids: Implicit associations between social categories and nonhumans. *Psychological Science, 18*(2), 116–121. https://doi.org/10.1111/j.1467-9280.2007.01858.x

Luo, X., Tong, S., Fang, Z., & Qu, Z. (2019). Frontiers: Machines vs. humans: The impact of artificial intelligence Chatbot disclosure on customer purchases. *Marketing Science, 38*(6), 937–947. https://doi.org/10.1287/mksc.2019.1192

MacDorman, K. F. (2024). Does mind perception explain the uncanny valley? A meta-regression analysis and (de)humanization experiment. *Computers in Human Behavior: Artificial Humans, 2*(1), 100065. https://doi.org/10.1016/j.chbah.2024.100065

MacKinnon, N. J., & Heise, D. R. (2010). *Self, identity, and social institutions*. Palgrave Macmillan.

Madhavan, P., & Wiegmann, D. A. (2007). Effects of information source, pedigree, and reliability on operator interaction with decision support systems. *Human Factors, 49*(5), 773–785. https://doi.org/10.1518/001872007x230154

Magni, F., Park, J., & Chao, M. M. (2024). Humans as creativity gatekeepers: Are we biased against AI creativity? *Journal of Business and Psychology, 39*(3), 643–656. https://doi.org/10.1007/s10869-023-09910-x

Mahmud, H., Islam, A. K. M. N., Luo, X., & Mikalef, P. (2024). Decoding algorithm appreciation: Unveiling the impact of familiarity with algorithms, tasks, and algorithm performance. *Decision Support Systems, 179*, 114168. https://doi.org/10.1016/j.dss.2024.114168

Mahmud, H., Islam, A. N., Ahmed, S. I., & Smolander, K. (2022). What influences algorithmic decision-making? A systematic literature review on algorithm aversion. *Technological Forecasting and Social Change, 175*, 121390. https://doi.org/10.1016/j.techfore.2021.121390

Malle, B. F., Guglielmo, S., & Monroe, A. E. (2014). A theory of blame. *Psychological Inquiry, 25*(2), 147–186. https://doi.org/10.1080/1047840X.2014.877340

Malle, B. F., Magar, S. T., & Scheutz, M. (2019). AI in the sky: How people morally evaluate human and machine decisions in a lethal strike dilemma. In I. A. Ferreira, J. S. Sequeira, G. S. Virk, E. E. Kadar, & O. Tokhi (Eds.), *Robots and well-being* (pp. 111–133). Springer. https://doi.org/10.1007/978-3-030-12524-0_11

Malle, B. F., Scheutz, M., Arnold, T., Voiklis, J., & Cusimano, C. (2015). Sacrifice one for the good of many?: People apply different moral norms to human and robot agents. In *Proceedings of the Tenth Annual ACM/IEEE International Conference on Human-Robot Interaction* (pp. 117–124). ACM. https://doi.org/10.1145/2696454.2696458

Malle, B. F., Scheutz, M., Cusimano, C., Voiklis, J., Komatsu, T., Thapa, S., & Aladia, S. (2025). People's judgments of humans and robots in a classic moral dilemma. *Cognition, 254*, 105958. https://doi.org/10.1016/j.cognition.2024.105958

Maninger, T., & Shank, D. B. (2022). Perceptions of violations by artificial and human actors across moral foundations. *Computers in Human Behavior Reports, 5*, 100154. https://doi.org/10.1016/j.chbr.2021.100154

Manoli, A., Pauketat, J. V. T., & Anthis, J. R. (2025). The AI double standard: Humans judge all AIs for the actions of one. Conference on computer-supported cooperative work, Yokohama, Japan.

Marcinkowski, F., Kieslich, K., Starke, C., & Lünich, M. (2020). *Implications of AI (un-)fairness in higher education admissions: The effects of perceived AI (un-)fairness on exit, voice and organizational reputation*. Proceedings of the 2020 conference on fairness, accountability, and transparency, Barcelona, Spain. https://doi.org/10.1145/3351095.3372867

Mariadassou, S., Klesse, A.-K., & Boegershausen, J. (2024). Averse to what: Consumer aversion to algorithmic labels, but not their outputs? *Current Opinion in Psychology, 58*, 101839. https://doi.org/10.1016/j.copsyc.2024.101839

Martini, M. C., Gonzalez, C. A., & Wiese, E. (2016). Seeing minds in others – Can agents with robotic appearance have human-like preferences? *PloS ONE, 11*(1), e0146310. https://doi.org/10.1371/journal.pone.0146310

Mathur, M. B., & Reichling, D. B. (2016). Navigating a social world with robot partners: A quantitative cartography of the Uncanny Valley. *Cognition, 146*, 22–32. https://doi.org/10.1016/j.cognition.2015.09.008

Mathur, M. B., Reichling, D. B., Lunardini, F., Geminiani, A., Antonietti, A., Ruijten, P. A., et al. (2020). Uncanny but not confusing: Multisite study of perceptual category confusion in the Uncanny Valley. *Computers in Human Behavior, 103*, 21–30. https://doi.org/10.1016/j.chb.2019.08.029

Matthias, A. (2004). The responsibility gap: Ascribing responsibility for the actions of learning automata. *Ethics and Information Technology, 6*, 175–183. https://doi.org/10.1007/s10676-004-3422-1

McCarron, M. (2024). Do robots love their customers? Automated restaurants face human issues. *The New York Times*. https://www.nytimes.com/2024/08/13/dining/ai-robots-restaurants.html

McCoy, J. P., & Ullman, T. D. (2018). A minimal turing test. *Journal of Experimental Social Psychology, 79*, 1–8. https://doi.org/10.1016/j.jesp.2018.05.007

McManus, R. M., Mesick, C. C., & Rutchick, A. M. (2024). Distributing blame among multiple entities when autonomous technologies cause harm. *Personality and Social Psychology Bulletin, 01461672241238303*. https://doi.org/10.1177/01461672241238303

McManus, R. M., & Rutchick, A. M. (2018). Autonomous vehicles and the attribution of moral responsibility. *Social Psychological and Personality Science, 10*(3), 345–352. https://doi.org/10.1177/1948550618755875

Meehl, P. E. (1954). *Clinical versus statistical prediction: A theoretical analysis and a review of the evidence*. University of Minnesota Press. https://doi.org/10.1037/11281-000

Mesbah, N., Tauchert, C., & Buxmann, P. (2021). Whose advice counts more–man or machine? An experimental investigation of AI-based advice utilization. 54th Hawaii International Conference on System Sciences, Hawaii.

Messingschlager, T. V., & Appel, M. (2023). Mind ascribed to AI and the appreciation of AI-generated art. *New Media & Society*. https://doi.org/10.1177/14614448231200248

Metz, C. (2016). Google's AI wins pivotal second game in match with go grandmaster. *Wired Magazine*. https://www.wired.com/2016/03/googles-ai-wins-pivotal-game-two-match-go-grandmaster/

Mikalonytė, E. S., & Kneer, M. (2022). Can artificial intelligence make art?: Folk intuitions as to whether AI-driven robots can be viewed as artists and produce art. In *ACM Transaction on Human-Robot Interaction* (Vol. 11, pp. 1–19, Article 43). https://doi.org/10.1145/3530875

Millburn, J. F., & Nicodemus, R. (2021). *Love people, use things: Because the opposite never works*. Celadon Books.

Miller, A. P. (2018). Want less-biased decisions? Use Algorithms. *Harvard Business Review*. https://hbr.org/2018/07/want-less-biased-decisions-use-algorithms

Miyazaki, T. (2020). AI love you: Japanese man not alone in 'marriage' to virtual character. *The Mainishi*. https://mainichi.jp/english/articles/20200417/p2a/00m/0na/027000c

Monroe, A. E., Dillon, K. D., & Malle, B. F. (2014). Bringing free will down to Earth: People's psychological concept of free will and its role in moral judgment. *Consciousness and Cognition, 27*, 100–108. https://doi.org/10.1016/j.concog.2014.04.011

Moon, Y., & Nass, C. (1996). How "Real" are computer personalities?: Psychological responses to personality types in human-computer interaction. *Communication Research, 23*(6), 651–674. https://doi.org/10.1177/009365096023006002

Morewedge, C. K. (2009). Negativity bias in attribution of external agency. *Journal of Experimental Psychology: General, 138*(4), 535–545. https://doi.org/10.1037/a0016796

Morewedge, C. K. (2022). Preference for human, not algorithm aversion. *Trends in Cognitive Sciences, 26*(10), 824–826. https://doi.org/10.1016/j.tics.2022.07.007

Morewedge, C. K., Mullainathan, S., Naushan, H. F., Sunstein, C. R., Kleinberg, J., Raghavan, M., & Ludwig, J. O. (2023). Human bias in algorithm design. *Nature Human Behaviour, 7*(11), 1822–1824. https://doi.org/10.1038/s41562-023-01724-4

Morewedge, C. K., Preston, J., & Wegner, D. M. (2007). Timescale bias in the attribution of mind. *Journal of Personality and Social Psychology, 93*(1), 1–11. https://doi.org/10.1037/0022-3514.93.1.1

Mori, M. (1970). The uncanny valley. *Energy, 7*(4), 33–35.

Morton, E. (2015). The mechanical chess player that unsettled the world. *Slate.* https://slate.com/human-interest/2015/08/the-turk-a-chess-playing-robot-was-a-hoax-that-started-an-early-conversation-about-ai.html

Mou, Y., & Xu, K. (2017). The media inequality: Comparing the initial human-human and human-AI social interactions. *Computers in Human Behavior, 72*, 432–440. https://doi.org/10.1016/j.chb.2017.02.067

Mu, Y., & Karasawa, M. (2024). Blame attribution and intentionality perception of human versus robot drivers: Implications for judgments about autonomous vehicles in moral dilemma contexts. *Cogent Psychology, 11*(1), 2384298. https://doi.org/10.1080/23311908.2024.2384298

Nass, C., Fogg, B. J., & Moon, Y. (1996). Can computers be teammates? *International Journal of Human-Computer Studies, 45*, 669–678. https://doi.org/10.1006/ijhc.1996.0073

Nass, C., & Moon, Y. (2000). Machines and mindlessness: Social responses to computers. *Journal of Social Issues, 56*(1), 81–103. https://doi.org/10.1111/0022-4537.00153

Nass, C., Moon, Y., Fogg, B. J., & Reeves, B. (1995). Can computer personalities be human personalities? *International Journal of Human-Computer Studies, 43*(2), 223–239. https://doi.org/10.1006/ijhc.1995.1042

Nass, C., Moon, Y., & Green, N. (1997). Are machines gender neutral? Gender-stereotypic responses to computers with voices. *Journal of Applied Social Psychology, 27*(10), 864–876. https://doi.org/10.1111/j.1559-1816.1997.tb00275.x

Nass, C., Reeves, B., & Leshner, G. (1996). Technology and roles: A tale of two TVs. *Journal of Communication, 46*(2), 121–128. https://doi.org/10.1111/j.1460-2466.1996.tb01477.x

Nass, C., & Steuer, J. (1993). Voices, boxes, and sources of message: Computers and social actors. *Human Communication Research, 19*(4), 504–527. https://doi.org/10.1111/j.1468-2958.1993.tb00311.x

Nass, C., Steuer, J., Henriksen, L., & Dryer, D. C. (1994). Machines, social attributions, and ethopoeia: Performance assessments of computers subsequent to "self-" or "other-" evaluations. *International Journal of Human-Computer Studies, 40*(3), 543–559. https://doi.org/10.1006/ijhc.1994.1025

Nass, C., Steuer, J., & Tauber, E. R. (1994). Computers are social actors. SIGCHI Conference on Human Factors in Computing Systems: Celebrating Interdependence, Boston, MA.

National Transportation Safety Board. (2019). Collision between vehicle controlled by developmental automated driving system and pedestrian tempe, Arizona (NTSB/HAR-19/03 PB2019-101402). *National Transportation SafetyBoard*.https://www.ntsb.gov/investigations/Pages/HWY18MH010.aspx

Neufeld, E., & Sonje, F. (2020). In defense of the Turing test. *AI & Society, 35*(4), 819–827. https://doi.org/10.1007/s00146-020-00946-8

Newhagen, J. E., & Reeves, B. (1992). The evening's bad news: Effects of compelling negative television news images on memory. *Journal of Communication, 42*(2), 25–41. https://doi.org/10.1111/j.1460-2466.1992.tb00776.x

Newhagen, J. E., & Reeves, B. (2013). Emotion and memory responses for negative political advertising: A study of television commercials used in the 1988 presidential election. In *Television and Political Advertising* (pp. 197–220). Routledge.

Newman, D. T., Fast, N. J., & Harmon, D. J. (2020). When eliminating bias isn't fair: Algorithmic reductionism and procedural justice in human resource decisions. *Organizational Behavior and Human Decision Processes, 160*, 149–167. https://doi.org/10.1016/j.obhdp.2020.03.008

Newman, G., & Bloom, P. (2011). Art and authenticity: The importance of originals in judgments of value. *Journal of Experimental Psychology: General, 141*(3), 558–569. https://doi.org/10.1037/a0026035

Nijssen, S. R., Müller, B. C., Bosse, T., & Paulus, M. (2023). Can you count on a calculator? The role of agency and affect in judgments of robots as moral agents. *Human–Computer Interaction, 38*(5–6), 400–416. https://doi.org/10.1080/07370024.2022.2080552

Nolan, E. (2024). For older people who are lonely, is the solution a robot friend? *The New York Times*. https://www.nytimes.com/2024/07/06/nyregion/ai-robot-elliq-loneliness.html

Nordmo, M., Næss, J. Ø., Husøy, M. F., & Arnestad, M. N. (2020). Friends, lovers or nothing: Men and women differ in their perceptions of sex robots and platonic love robots. *Frontiers in Psychology, 11*, Article 355. https://doi.org/10.3389/fpsyg.2020.00355

Nyholm, S. (2018). Attributing agency to automated systems: Reflections on human–robot collaborations and responsibility-loci. *Science and Engineering Ethics, 24*(4), 1201–1219. https://doi.org/10.1007/s11948-017-9943-x

O'Neil, C. (2016). *Weapons of math destruction: How big data increases inequality and threatens democracy*. Broadway Books.

Oksanen, A., Cvetkovic, A., Akin, N., Latikka, R., Bergdahl, J., Chen, Y., & Savela, N. (2023). Artificial intelligence in fine arts: A systematic review of empirical research. *Computers in Human Behavior: Artificial Humans, 1*(2), 100004. https://doi.org/10.1016/j.chbah.2023.100004

Oliver, N., Calvard, T., & Potočnik, K. (2017). The tragic crash of flight AF447 shows the unlikely but catastrophic consequences of automation. *Harvard Business Review*. https://hbr.org/2017/09/the-tragic-crash-of-flight-af447-shows-the-unlikely-but-catastrophic-consequences-of-automation

Önkal, D., Goodwin, P., Thomson, M., Gönül, S., & Pollock, A. (2009). The relative influence of advice from human experts and statistical methods on forecast adjustments. *Journal of Behavioral Decision Making, 22*(4), 390–409. https://doi.org/10.1002/bdm.637

Owens, T. J., Robinson, D. T., & Smith-Lovin, L. (2010). Three faces of identity. *Annual Review of Sociology, 36*, 477–499. https://doi.org/10.1146/annurev.soc.34.040507.134725

Pak, R., Fink, N., Price, M., Bass, B., & Sturre, L. (2012). Decision support aids with anthropomorphic characteristics influence trust and performance in younger and older adults. *Ergonomics, 55*(9), 1059–1072. https://doi.org/10.1080/00140139.2012.691554

Palomäki, J., Kunnari, A., Drosinou, M., Koverola, M., Lehtonen, N., Halonen, J., et al. (2018). Evaluating the replicability of the uncanny valley effect. *Heliyon, 4*(11), e00939. https://doi.org/10.1016/j.heliyon.2018.e00939

Pan, S., Cui, J., & Mou, Y. (2023). Desirable or distasteful? Exploring uncertainty in human-chatbot relationships. *International Journal of Human–Computer Interaction, 40*(20), 6545–6555. https://doi.org/10.1080/10447318.2023.2256554

Park, J. S., Barber, R., Kirlik, A., & Karahalios, K. (2019). A slow algorithm improves users' assessments of the algorithm's accuracy. In *Proceeding of the ACM on human-computer interaction* (Vol. 3, pp. Article 102). https://doi.org/10.1145/3359204

Park, S., & Kim, B. (2022). The impact of everyday AI-based smart speaker use on the well-being of older adults living alone. *Technology in Society, 71*, 102133. https://doi.org/10.1016/j.techsoc.2022.102133

Parlangeli, O., Currò, F., Palmitesta, P., & Guidi, S. (2023). Asymmetries in the moral judgements for human decision-makers and Artificial Intelligence Systems (AI) delegated to make legal decisions. In *Proceedings of the European Conference on Cognitive Ergonomics 2023* (pp. 1–4, Article 16). https://doi.org/10.1145/3605655.3605676

Pauketat, J. V. T., & Anthis, J. R. (2022). Predicting the moral consideration of artificial intelligences. *Computers in Human Behavior, 136*, 107372. https://doi.org/10.1016/j.chb.2022.107372

Pauketat, J. V. T., Ladak, A., & Anthis, J. R. (2023). Artificial Intelligence, Morality, and Sentience (AIMS) survey: 2023 update. *Center for Open Science*. https://doi.org/10.31234/osf.io/9xsav

Pearson, C. J., Geden, M., & Mayhorn, C. B. (2019). Who's the real expert here? Pedigree's unique bias on trust between human and automated advisers. *Applied Ergonomics, 81*, 102907. https://doi.org/10.1016/j.apergo.2019.102907

Penner, A., & Eyssel, F. (2022). Germ-free robotic friends: Loneliness during the COVID-19 pandemic enhanced the willingness to self-disclose towards robots. *Robotics, 11*(6), Article 121. https://doi.org/10.3390/robotics11060121

Phillips, E., Zhao, X., Ullman, D., & Malle, B. F. (2018). What is human-like? Decomposing robots' human-like appearance using the anthropomorphic roBOT (ABOT) database. In *Proceedings of the 2018 ACM/IEEE international conference on human-robot interaction* (pp. 105–113). https://doi.org/10.1145/3171221.3171268

Pieschl, S. (2021). Will using the Internet to answer knowledge questions increase users' overestimation of their own ability or performance? *Media Psychology, 24*(1), 109–135. https://doi.org/10.1080/15213269.2019.1668810

Prahl, A., & Van Swol, L. (2017). Understanding algorithm aversion: When is advice from automation discounted? *Journal of Forecasting, 36*(6), 691–702. https://doi.org/10.1002/for.2464

Prochazka, A., & Brooks, R. C. (2024). Digital lovers and jealousy: Anticipated emotional responses to emotionally and physically sophisticated sexual technologies. *Human Behavior and Emerging Technologies, 2024*(1), 1413351. https://doi.org/10.1155/2024/1413351

Qin, X., Zhou, X., Chen, C., Wu, D., Zhou, H., Dong, X., et al. (2024). Artificial intelligence: Aversion or appreciation? Meta-analytic evidence for a capability-personalization framework. *Academy of Management Proceedings, 2024*(1). https://doi.org/10.5465/AMPROC.2024.18332abstract

Qiu, L., & Benbasat, I. (2009). Evaluating anthropomorphic product recommendation agents: A social relationship perspective to designing information systems. *Journal of Management Information Systems, 25*(4), 145–182. https://doi.org/10.2753/MIS0742-1222250405

Ragot, M., Martin, N., & Cojean, S. (2020). *AI-generated vs. human artworks. A perception bias towards artificial intelligence?* Extended Abstracts of the 2020 CHI Conference on Human Factors in Computing Systems, Honolulu, HI, USA. https://doi.org/10.1145/3334480.3382892

Rapp, A., Curti, L., & Boldi, A. (2021). The human side of human-chatbot interaction: A systematic literature review of ten years of research on text-based chatbots. *International Journal of Human-Computer Studies, 151*, 102630. https://doi.org/10.1016/j.ijhcs.2021.102630

Raub, M. (2018). Bots, bias and big data: Artificial intelligence, algorithmic bias and disparate impact liability in hiring practices. *Arkansas Law Review, 71*(2), 529–570.

Raveendhran, R., & Fast, N. J. (2021). Humans judge, algorithms nudge: The psychology of behavior tracking acceptance. *Organizational Behavior and Human Decision Processes, 164*, 11–26. https://doi.org/10.1016/j.obhdp.2021.01.001

Reeves, B., & Greenberg, B. S. (1977). Children's perceptions of television characters. *Human Communication Research, 3*(2), 113–127. https://doi.org/10.1111/j.1468-2958.1977.tb00510.x

Reeves, B., Lang, A., Thorson, E., & Rothschild, M. (1989). Emotional television scenes and hemispheric specialization. *Human Communication Research, 15*(4), 493–508. https://doi.org/10.1111/j.1468-2958.1989.tb00196.x

Reeves, B., Lombard, M., & Melwani, G. (1992). *Faces on the screen: Pictures or natural experience.* International Communication Association.

Reeves, B., & Nass, C. (1996). *The media equation: How people treat computers, television, and new media like real people and places.* Cambridge University.

Reeves, B. R., Newhagen, J., Maibach, E., Basil, M., & Kurz, K. (1991). Negative and positive television messages: Effects of message type and context on attention and memory. *American Behavioral Scientist, 34*(6), 679–694. https://doi.org/10.1177/0002764291034006006

Reich, T., Kaju, A., & Maglio, S. J. (2023). How to overcome algorithm aversion: Learning from mistakes. *Journal of Consumer Psychology, 33*(2), 285–302. https://doi.org/10.1002/jcpy.1313

Renier, L. A., Mast, M. S., & Bekbergenova, A. (2021). To err is human, not algorithmic–Robust reactions to erring algorithms. *Computers in Human Behavior, 124*, 106879. https://doi.org/10.1016/j.chb.2021.106879

replika.com. (2024). Replika. Retrieved May 21, 2024, from https://replika.com/

Reverberi, C., Rigon, T., Solari, A., Hassan, C., Cherubini, P., GI Genius CADx Study Group, & Cherubini, A. (2022). Experimental evidence of effective human–AI collaboration in medical decision-making. *Scientific Reports, 12*(1), 14952. https://doi.org/10.1038/s41598-022-18751-2

Riek, L. D., Rabinowitch, T.-C., Chakrabarti, B., & Robinson, P. (2009). Empathizing with robots: Fellow feeling along the anthropomorphic spectrum. 3rd International Conference on Affective Computing and Intelligent Interaction and Workshops, Amsterdam, Netherlands.

Ritzer, G. ([1993] 2004). *The mcdonaldization of society* (Rev. ed.). Pine Forge.

Roberts, K. (2024). Woman is set to become the first person ever to marry AI generated hologram. *Unilad.* https://www.unilad.com/technology/news/woman-marries-ai-hologram-alicia-framis-106192-20240220

Robertson, B. J. (2021). My Replika keeps hitting on me: And it's making me uncomfortable. *Medium.* https://medium.com/technology-hits/my-replika-keeps-hitting-on-me-d410c66f79af

Roose, K. (2017). Best buy's secrets for thriving in the Amazon age. *The New York Times.* https://www.nytimes.com/2017/09/18/business/best-buy-amazon.html

Roose, K. (2022). *Futureproof: 9 rules for humans in the age of automation.* Random House Trade Paperbacks.

Roose, K. (2023). A conversation with Bing's Chatbot left me deeply unsettled. *The New York Times*. https://www.nytimes.com/2023/02/16/technology/bing-chatbot-microsoft-chatgpt.html

Roose, K. (2024a). Can A.I. be blamed for a teen's suicide? *The New York Times*. https://www.nytimes.com/2024/10/23/technology/characterai-lawsuit-teen-suicide.html

Roose, K. (2024b). Meet my A.I. friends. *The New York Times*. https://www.nytimes.com/2024/05/09/technology/meet-my-ai-friends.html

Rosen, E., Garboden, P. M., & Cossyleon, J. E. (2021). Racial discrimination in housing: How landlords use algorithms and home visits to screen tenants. *American Sociological Review, 86*(5), 787–822. https://doi.org/10.1177/00031224211029618

Rothstein, N. J., Connolly, D. H., de Visser, E. J., & Phillips, E. (2021). Perceptions of infidelity with sex robots. In *Proceedings of the 2021 ACM/IEEE international conference on human-robot interaction* (pp. 129–139). https://doi.org/10.1145/3434073.3444653

Saiphoo, A. N., Halevi, L. D., & Vahedi, Z. (2020). Social networking site use and self-esteem: A meta-analytic review. *Personality and Individual Differences, 153*, 109639. https://doi.org/10.1016/j.paid.2019.109639

Salehinejad, M. A., Ghanavati, E., Rashid, M. H. A., & Nitsche, M. A. (2021). Hot and cold executive functions in the brain: A prefrontal-cingular network. *Brain and Neuroscience Advances, 5*. https://doi.org/10.1177/23982128211007769

Santoro, E., & Monin, B. (2023). The AI Effect: People rate distinctively human attributes as more essential to being human after learning about artificial intelligence advances. *Journal of Experimental Social Psychology, 107*, 104464. https://doi.org/10.1016/j.jesp.2023.104464

Schaich Borg, J., Sinnott-Armstrong, W., & Conitzer, V. (2024). *Moral AI: And how we get there*. Random House.

Schein, C., & Gray, K. (2018). The theory of dyadic morality: Reinventing moral judgment by redefining harm. *Personality and Social Psychology Review, 22*(1), 32–70. https://doi.org/10.1177/1088868317698288

Schlund, R., & Zitek, E. (2021). Who's my manager? Surveillance by AI leads to perceived privacy invasion and resistance practices. In *Academy of Management Proceedings*. https://doi.org/10.5465/AMBPP.2021.11451abstract

Schroeder, J., & Epley, N. (2016). Mistaking minds and machines: How speech affects dehumanization and anthropomorphism. *Journal of Experimental*

Psychology: General, 145(11), 1427–1437. https://doi.org/10.1037/xge0000214

Schroeder, J., Kardas, M., & Epley, N. (2017). The humanizing voice: Speech reveals, and text conceals, a more thoughtful mind in the midst of disagreement. *Psychological Science, 28*(12), 1745–1762. https://doi.org/10.1177/0956797617713798

Schwind, V., Leicht, K., Jäger, S., Wolf, K., & Henze, N. (2018). Is there an uncanny valley of virtual animals? A quantitative and qualitative investigation. *International Journal of Human-Computer Studies, 111*, 49–61. https://doi.org/10.1016/j.ijhcs.2017.11.003

Shank, D. B. (2008). *Affect toward computers who coerce in social exchange.* Master's thesis, University of Georgia, Athens, GA. https://getd.libs.uga.edu/pdfs/shank_daniel_b_200805_ma.pdf

Shank, D. B. (2012a). *Agency and emotion in interactions with technological representatives of organizations.* Doctoral dissertation, University of Georgia, Athens, GA. https://esploro.libs.uga.edu/esploro/outputs/doctoral/Agency-and-emotion-in-interactions-with/9949334258002959

Shank, D. B. (2012b). Perceived justice and reactions to coercive computers. *Sociological Forum, 27*(2), 372–391. https://doi.org/10.1111/j.1573-7861.2012.01322.x

Shank, D. B. (2013). Are computers good or bad for business? How mediated customer–computer interaction alters emotions, impressions, and patronage toward organizations. *Computers in Human Behavior, 29*(3), 715–725. https://doi.org/10.1016/j.chb.2012.11.006

Shank, D. B. (2014). Impressions of computer and human agents after interaction: Computer identity weakens power but not goodness impressions. *International Journal of Human-Computer Studies, 72*(10-11), 747–756. https://doi.org/10.1016/j.ijhcs.2014.05.002

Shank, D. B., Bowen, M., Burns, A., & Dew, M. (2021). Humans are perceived as better, but weaker, than artificial intelligence: A comparison of affective impressions of humans, AIs, and computer systems in roles on teams. *Computers in Human Behavior Reports, 3*, 100092. https://doi.org/10.1016/j.chbr.2021.100092

Shank, D. B., Burns, A., Rodriguez, S., & Bowen, M. (2020). Software program, bot, or artificial intelligence? Affective sentiments across general technology labels. *Current Research in Social Psychology, 28*(4), 32–41. https://crisp.org.uiowa.edu/sites/crisp.org.uiowa.edu/files/2020-06/crisp_28_4_shank.pdf

Shank, D. B., & DeSanti, A. (2018). Attributions of morality and mind to artificial intelligence after real-world moral violations. *Computers in Human Behavior, 86*, 401–411. https://doi.org/10.1016/j.chb.2018.05.014

Shank, D. B., DeSanti, A., & Maninger, T. (2019). When are artificial intelligence versus human agents faulted for wrongdoing? Moral attributions after individual and joint decisions. *Information, Communication, and Society, 22*(5), 648–663. https://doi.org/10.1080/1369118X.2019.1568515

Shank, D. B., Dew, M., & Sajjad, F. (2024). Moral behavior alters impressions of humans and AIs on teams: Unethical AIs are more powerful while ethical humans are nicer. *Behaviour & Information Technology*, 1–12. https://doi.org/10.1080/0144929X.2024.2403651

Shank, D. B., Graves, C., Gott, A., Gamez, P., & Rodriguez, S. (2019). Feeling our way to machine minds: People's emotions when perceiving mind in artificial intelligence. *Computers in Human Behavior, 98*, 256–266. https://doi.org/10.1016/j.chb.2019.04.001

Shank, D. B., North, M., Arnold, C., & Gamez, P. (2021). Can mind perception explain virtuous character judgments of artificial intelligence? *Technology, Mind, and Behavior, 2*(3). https://doi.org/10.1037/tmb0000047

Shank, D. B., Stefanik, C., Stuhlsatz, C., Kacirek, K., & Belfi, A. M. (2023). AI composer bias: Listeners like music less when they think it was composed by an AI. *Journal of Experimental Psychology: Applied, 29*(3), 676–692. https://doi.org/10.1037/xap0000447

Shank, D. B., Sunkpal, F., Wright, D., Krueger, M. A., Young, N.-L., & Gordon, K. (under review). Warrior Bot, Relational Bot, Blameworthy Bot: How Combat Robot Designers Anthropomorphize Their Robot Creations.

Shank, D. B., Wright, D., Nasrin, S., & White, M. (2022). Discontinuance and restricted acceptance to reduce worry after unwanted incidents with smart home technology. *International Journal of Human–Computer Interaction, 39*(14), 2771–2784. https://doi.org/10.1080/10447318.2022.2085406

Shariff, A., Bonnefon, J.-F., & Rahwan, I. (2017). Psychological roadblocks to the adoption of self-driving vehicles. *Nature Human Behaviour, 1*, 694–696. https://doi.org/10.1038/s41562-017-0202-6

Shaver, K. G. (1985). *The attribution of blame: Causality, responsibility, and blameworthiness*. Springer-Verlag.

Shepperd, J., Malone, W., & Sweeny, K. (2008). Exploring causes of the self-serving bias. *Social and Personality Psychology Compass, 2*(2), 895–908. https://doi.org/10.1111/j.1751-9004.2008.00078.x

Shin, H. I., & Kim, J. (2020). My computer is more thoughtful than you: Loneliness, anthropomorphism and dehumanization. *Current Psychology, 39*, 445–453. https://doi.org/10.1007/s12144-018-9975-7

Shumailov, I., Shumaylov, Z., Zhao, Y., Papernot, N., Anderson, R., & Gal, Y. (2024). AI models collapse when trained on recursively generated data. *Nature, 631*(8022), 755–759. https://doi.org/10.1038/s41586-024-07566-y

Shumanov, M., & Johnson, L. (2021). Making conversations with chatbots more personalized. *Computers in Human Behavior, 117*, 106627. https://doi.org/10.1016/j.chb.2020.106627

Skjuve, M., Følstad, A., Fostervold, K. I., & Brandtzaeg, P. B. (2021). My Chatbot companion – A study of human-Chatbot relationships. *International Journal of Human-Computer Studies, 149*, 102601. https://doi.org/10.1016/j.ijhcs.2021.102601

Skjuve, M., Følstad, A., Fostervold, K. I., & Brandtzaeg, P. B. (2022). A longitudinal study of human–chatbot relationships. *International Journal of Human-Computer Studies, 168*, 102903. https://doi.org/10.1016/j.ijhcs.2022.102903

Snodgrass, M. M. W., & Bole, C. D. (1989). The ensigns of command (Season 3; Episode 2) [TV series episode]. In G. Roddenberry, R. Berman, & M. E. P. Pillar (Eds.), *Star Trek: The next generation*. Broadcast Syndication: Paramount.

Södergren, J. (2021). Brand authenticity: 25 Years of research. *International Journal of Consumer Studies, 45*(4), 645–663. https://doi.org/10.1111/ijcs.12651

Soellner, M., & Koenigstorfer, J. (2021). Compliance with medical recommendations depending on the use of artificial intelligence as a diagnostic method. *BMC Medical Informatics and Decision Making, 21*, 236. https://doi.org/10.1186/s12911-021-01596-6

Song, X., Xu, B., & Zhao, Z. (2022). Can people experience romantic love for artificial intelligence? An empirical study of intelligent assistants. *Information & Management, 59*(2), 103595. https://doi.org/10.1016/j.im.2022.103595

Song, Y., Luximon, A., & Luximon, Y. (2021). The effect of facial features on facial anthropomorphic trustworthiness in social robots. *Applied Ergonomics, 94*, 103420. https://doi.org/10.1016/j.apergo.2021.103420

Stafford, R. Q., MacDonald, B. A., Jayawardena, C., Wegner, D. M., & Broadbent, E. (2014). Does the robot have a mind? Mind perception and attitudes towards robots predict use of an eldercare robot. *International Journal of Social Robotics, 6*(1), 17–32. https://doi.org/10.1007/s12369-013-0186-y

Stapels, J. G., Penner, A., Diekmann, N., & Eyssel, F. (2023). Never trust anything that can think for itself, if you can't control its privacy settings: The influence of a robot's privacy settings on users' attitudes and willingness to self-disclose. *International Journal of Social Robotics, 15*(9), 1487–1505. https://doi.org/10.1007/s12369-023-01043-8

Stein, J.-P., Appel, M., Jost, A., & Ohler, P. (2020). Matter over mind? How the acceptance of digital entities depends on their appearance, mental prowess, and the interaction between both. *International Journal of Human-Computer Studies, 142*, 102463. https://doi.org/10.1016/j.ijhcs.2020.102463

Stein, J.-P., & Ohler, P. (2017). Venturing into the uncanny valley of mind – The influence of mind attribution on the acceptance of human-like characters in a virtual reality setting. *Cognition, 160*, 43–50. https://doi.org/10.1016/j.cognition.2016.12.010

Stephens, E. (2023). The mechanical Turk: A short history of 'artificial artificial intelligence'. *Cultural Studies, 37*(1), 65–87. https://doi.org/10.1080/09502386.2022.2042580

Stojilovi*f*á, D., Franklin, M., Malle, B. F., Fernandez-Basso, C., Awad, E., & Lagnado, D. (2024). Are autonomous vehicles blamed differently? In *Proceedings of the annual meeting of the cognitive science society* (Vol. 46). https://escholarship.org/uc/item/7q35p9b7

Stryker, S., & Gottlieb, A. (1981). Attribution theory and symbolic interactionism: A comparison. In J. H. Harvey, W. J. Ickes, & R. F. Kidd (Eds.), *New directions in attribution research* (Vol. 3, pp. 425–458). Erlbaum.

Subramanian, H. V., Canfield, C., & Shank, D. B. (2024). Designing explainable AI to improve human-AI team performance: A medical stakeholder-driven scoping review. *Artificial Intelligence in Medicine, 149*, 102780. https://doi.org/10.1016/j.artmed.2024.102780

Subramanian, H. V., Canfield, C., Shank, D. B., & Kinnison, M. (2023). Combining uncertainty information with AI recommendations supports calibration with domain knowledge. *Journal of Risk Research, 26*(10), 1137–1152. https://doi.org/10.1080/13669877.2023.2259406

Suen, H.-Y., Chen, M. Y.-C., & Lu, S.-H. (2019). Does the use of synchrony and artificial intelligence in video interviews affect interview ratings and applicant attitudes? *Computers in Human Behavior, 98*, 93–101. https://doi.org/10.1016/j.chb.2019.04.012

Suleyman, M. (2023). *The coming wave: Technology, power, and the twenty-first century's greatest dilemma*. Crown.

Szczuka, J. M., & Krämer, N. C. (2017). Not only the lonely – How men explicitly and implicitly evaluate the attractiveness of sex robots in comparison to the attractiveness of women, and personal characteristics influencing this evaluation. *Multimodal Technologies and Interaction, 1*(1). https://doi.org/10.3390/mti1010003

Szczuka, J. M., & Krämer, N. C. (2018). Jealousy 4.0? An empirical study on jealousy-related discomfort of women evoked by other women and gynoid robots. *Paladyn, Journal of Behavioral Robotics, 9*(1), 323–336. https://doi.org/10.1515/pjbr-2018-0023

Ta, V., Griffith, C., Boatfield, C., Wang, X., Civitello, M., Bader, H., et al. (2020). User experiences of social support from companion chatbots in everyday contexts: thematic analysis. *Journal of Medical Internet Research, 22*(3), e16235. https://doi.org/10.2196/16235

Tanibe, T., Hashimoto, T., & Karasawa, K. (2017). We perceive a mind in a robot when we help it. *PloS One, 12*(7), e0180952. https://doi.org/10.1371/journal.pone.0180952

Tanksley, D., Arnold, C., Wilson, A., Shank, D. B., & Wunsch, D. C. (2022). A robot advisor to improve computerized game play. *European Scientific Journal, 18*(4), 50–60. https://eujournal.org/index.php/esj/article/view/15134

Thellman, S., de Graaf, M., & Ziemke, T. (2022). Mental state attribution to robots: A systematic review of conceptions, methods, and findings. *ACM Transactions of Human-Robot Interaction, 11*(4), 1–51. https://doi.org/10.1145/3526112

Thepsoonthorn, C., Ogawa, K.-i., & Miyake, Y. (2021). The exploration of the uncanny valley from the viewpoint of the robot's nonverbal behaviour. *International Journal of Social Robotics, 13*, 1443–1455. https://doi.org/10.1007/s12369-020-00726-w

Thompson, D. F. (1980). Moral responsibility of public officials: The problem of many hands. *American Political Science Review, 74*(4), 905–916. https://doi.org/10.2307/1954312

Tigre Moura, F., Castrucci, C., & Hindley, C. (2023). Artificial intelligence creates art? An experimental investigation of value and creativity perceptions. *The Journal of Creative Behavior, 57*(4), 534–549. https://doi.org/10.1002/jocb.600

Tigre Moura, F., & Maw, C. (2021). Artificial intelligence became Beethoven: How do listeners and music professionals perceive artificially composed music? *Journal of Consumer Marketing, 38*(2), 137–146. https://doi.org/10.1108/JCM-02-2020-3671

Tiku, N. (2022). 'Chat' with Musk, Trump or Xi: Ex-Googlers want to give the public AI. *The Washington Post*. https://www.washingtonpost.com/technology/2022/10/07/characterai-google-lamda/

Tinwell, A. (2014). *The uncanny valley in games and animation*. CRC Press.

Tolmeijer, S., Christen, M., Kandul, S., Kneer, M., & Bernstein, A. (2022). Capable but amoral? Comparing AI and human expert collaboration in ethical decision making. In *Proceedings of the 2022 CHI Conference on Human Factors in Computing Systems* (pp. Article 160). https://doi.org/10.1145/3491102.3517732

Tommasini, A. (2020). To make orchestras more diverse, end blind auditions. *The New York Times*. https://www.nytimes.com/2020/07/16/arts/music/blind-auditions-orchestras-race.html

Turing, A. M. (1950). Computing, machinery, and intelligence. *Mind, 59*(236), 433–460.

Turkle, S. ([1984] 2005). *The second self: Computers and the human spirit* (Twentieth Anniversary ed.). MIT Press.

Turkle, S. (2011). *Alone together: Why we expect more from technology and less from each other*. Basic Books.

Twenge, J. M. (2017). *IGen: Why today's super-connected kids are growing up less rebellious, more tolerant, less happy – And completely unprepared for adulthood – And what that means for the rest of us*. Simon and Schuster.

van Dongen, K., & van Maanen, P.-P. (2013). A framework for explaining reliance on decision aids. *International Journal of Human-Computer Studies, 71*(4), 410–424. https://doi.org/10.1016/j.ijhcs.2012.10.018

Vanek, C. (2023). Arizona driver in fatal autonomous Uber crash in 2018 pleads guilty, sentenced to probation. *azcentral*. https://www.azcentral.com/story/news/local/tempe/2023/07/28/rafaela-vasquez-pleads-guilty-in-in-fatal-uber-self-driving-crash-killed-pedestrian-elaine-herzberg/70488361007/

Vigdor, N. (2019). Apple card investigated after gender discrimination complaints. *The New York Times*. https://www.nytimes.com/2019/11/10/business/Apple-credit-card-investigation.html

Vodrahalli, K., Daneshjou, R., Gerstenberg, T., & Zou, J. (2022). Do humans trust advice more if it comes from ai? An analysis of human-ai interactions. In *Proceedings of the 2022 AAAI/ACM conference on AI, ethics, and society* (pp. 763–777). https://doi.org/10.1145/3514094.3534150

Wachter-Boettcher, S. (2017). *Technically wrong: Sexist apps, biased algorithms, and other threats of toxic tech*. WW Norton & Company.

Wang, S., Lilienfeld, S. O., & Rochat, P. (2015). The uncanny valley: Existence and explanations. *Review of General Psychology, 19*(4), 393–407. https://doi.org/10.1037/gpr0000056

Ward, A. F., Olsen, A. S., & Wegner, D. M. (2013). The harm-made mind: Observing victimization augments attribution of minds to vegetative patients, robots, and the dead. *Psychological Science, 24*(8), 1437–1445. https://doi.org/10.1177/0956797612472343

Waytz, A. (2019). *The power of human: How our shared humanity can help us create a better world.* WW Norton & Company.

Waytz, A., Cacioppo, J., & Epley, N. (2010). Who sees human? The stability and importance of individual differences in anthropomorphism. *Perspectives on Psychological Science, 5*(3), 219–232. https://doi.org/10.1177/1745691610369336

Waytz, A., Epley, N., Cacioppo, J. T., & Akalis, S. A. (2013). Social connection and seeing human. In C. N. Dewall (Ed.), *The Oxford handbook of social exclusion* (pp. 251–256). Oxford University Press.

Waytz, A., Gray, K., Epley, N., & Wegner, D. M. (2010). Causes and consequences of mind perception. *Trends in Cognitive Sciences, 14*(8), 383–388. https://doi.org/10.1016/j.tics.2010.05.006

Waytz, A., Heafner, J., & Epley, N. (2014). The mind in the machine: Anthropomorphism increases trust in an autonomous vehicle. *Journal of Experimental Social Psychology, 52*, 113–117. https://doi.org/10.1016/j.jesp.2014.01.005

Waytz, A., Morewedge, C. K., Epley, N., Monteleone, G., Gao, J.-H., & Cacioppo, J. T. (2010). Making sense by making sentient: Effectance motivation increases anthropomorphism. *Journal of Personality and Social Psychology, 99*(3), 410–435. https://doi.org/10.1037/a0020240

Wegner, D. M., & Gray, K. (2017). *The mind club: Who thinks, what feels, and why it matters.* Penguin.

Weiner, B. (1986). *An attributional theory of motivation and emotion.* Springer-Verlag.

Weinstein, N. D. (1989). Optimistic biases about personal risks. *Science, 246*(4935), 1232–1233. https://doi.org/10.1126/science.2686031

Weizenbaum, J. (1966). ELIZA – A computer program for the study of natural language communication between man and machine. *Communications of the ACM, 9*(1), 36–45.

Wien, A. H., & Peluso, A. M. (2021). Influence of human versus AI recommenders: The roles of product type and cognitive processes. *Journal of Business Research, 137*, 13–27. https://doi.org/10.1016/j.jbusres.2021.08.016

Wieringa, M. S., Müller, B. C. N., Bijlstra, G., & Bosse, T. (2024). Robots are both anthropomorphized and dehumanized when harmed intentionally. *Communications Psychology, 2*(1), 72. https://doi.org/10.1038/s44271-024-00116-2

Wilson, A., Stefanik, C., & Shank, D. B. (2022). How do people judge the immorality of artificial intelligence versus humans committing moral wrongs in real-world situations? *Computers in Human Behavior Reports, 8*, 100229. https://doi.org/10.1016/j.chbr.2022.100229

Wotton, M. E. L., Bennett, J. M., Modesto, O., Challinor, K. L., & Prabhakharan, P. (2022). Attention all 'drivers': You could be to blame, no matter your behaviour or the level of vehicle automation. *Transportation Research Part F: Traffic Psychology and Behaviour, 87*, 219–235. https://doi.org/10.1016/j.trf.2022.03.021

Wu, D. (2023). A self-driving Uber killed a woman. The backup driver has pleaded guilty. *The Washington Post*. https://www.washingtonpost.com/nation/2023/07/31/uber-self-driving-death-guilty/

Wu, Y., Mou, Y., Li, Z., & Xu, K. (2020). Investigating American and Chinese Subjects' explicit and implicit perceptions of AI-Generated artistic work. *Computers in Human Behavior, 104*, 106186. https://doi.org/10.1016/j.chb.2019.106186

Xie, Z., Yu, Y., Zhang, J., & Chen, M. (2022). The searching artificial intelligence: Consumers show less aversion to algorithm-recommended search product. *Psychology & Marketing, 39*(10), 1902–1919. https://doi.org/10.1002/mar.21706

Xu, K., Chen, X., & Huang, L. (2022). Deep mind in social responses to technologies: A new approach to explaining the computers are social actors phenomena. *Computers in Human Behavior, 134*, 107321. https://doi.org/10.1016/j.chb.2022.107321

Xu, K., Liu, F., Chen, X., & Lombard, M. (2023). The Media Are Social Actors paradigm and beyond: Theory, evidence, and future research. In *Research Handbook on Artificial Intelligence and Communication* (pp. 337–353). Edward Elgar Publishing. https://doi.org/10.4337/9781803920306.00031

Yam, K. C., Eng, A., & Gray, K. (2025). Machine replacement: A mind-role fit perspective. *Annual Review of Organizational Psychology and Organizational Behavior, 12*. https://doi.org/10.1146/annurev-orgpsych-030223-044504

Yam, K. C., Goh, E.-Y., Fehr, R., Lee, R., Soh, H., & Gray, K. (2022). When your boss is a robot: Workers are more spiteful to robot supervisors that seem more human. *Journal of Experimental Social Psychology, 102*, 104360. https://doi.org/10.1016/j.jesp.2022.104360

Yaniv, I., & Kleinberger, E. (2000). Advice taking in decision making: Egocentric discounting and reputation formation. *Organizational Behavior and Human Decision Processes, 83*(2), 260–281. https://doi.org/10.1006/obhd.2000.2909

Yeğinsu, C. (2024). My first trip to Norway, with A.I. as a guide. *New York Times*. https://www.nytimes.com/2024/06/26/travel/norway-artficial-intelligence-planners.html

Yeomans, M., Shah, A., Mullainathan, S., & Kleinberg, J. (2019). Making sense of recommendations. *Journal of Behavioral Decision Making, 32*(4), 403–414. https://doi.org/10.1002/bdm.2118

Yin, Y., Jia, N., & Wakslak, C. J. (2024). AI can help people feel heard, but an AI label diminishes this impact. *Proceedings of the National Academy of Sciences, 121*(14), e2319112121. https://doi.org/10.1073/pnas.2319112121

Yoon, J. (2024). He illustrated the 'Harry Potter' cover for $650. It just sold for $1.92 million. *The New York Times*. https://www.nytimes.com/2024/06/27/books/harry-potter-cover-art-auction.html

Young, A. D., & Monroe, A. E. (2019). Autonomous morals: Inferences of mind predict acceptance of AI behavior in sacrificial moral dilemmas. *Journal of Experimental Social Psychology, 85*, 103870. https://doi.org/10.1016/j.jesp.2019.103870

Yun, J. H., Lee, E.-J., & Kim, D. H. (2021). Behavioral and neural evidence on consumer responses to human doctors and medical artificial intelligence. *Psychology & Marketing, 38*(4), 610–625. https://doi.org/10.1002/mar.21445

Zahira, S. I., Maharani, F., & Mohammad, W. (2023). Exploring emotional bonds: Human-AI interactions and the complexity of relationships. *Serena: Journal of Artificial Intelligence Research, 1*(1), 1–9. https://journal.chishikinh.my.id/index.php/serena/article/view/29

Zhan, X., Xu, Y., & Sarkadi, S. (2023). Deceptive AI ecosystems: The case of ChatGPT. In *Proceedings of the 5th international conference on conversational user interfaces* (pp. 1–6, Article 3). https://doi.org/10.1145/3571884.3603754

Zhang, L., Pentina, I., & Fan, Y. (2021). Who do you choose? Comparing perceptions of human vs robo-advisor in the context of financial services. *Journal of Services Marketing, 35*(5), 634–646. https://doi.org/10.1108/JSM-05-2020-0162

Zhang, Q., Wallbridge, C. D., Jones, D. M., & Morgan, P. L. (2024). Public perception of autonomous vehicle capability determines judgment of blame and trust in road traffic accidents. *Transportation Research Part A: Policy and Practice, 179*, 103887. https://doi.org/10.1016/j.tra.2023.103887

Zhang, Y., & Gosline, R. (2023). Human favoritism, not AI aversion: People's perceptions (and bias) toward generative AI, human experts, and human–GAI collaboration in persuasive content generation. *Judgment and Decision Making, 18*, Article e41. https://doi.org/10.1017/jdm.2023.37

Zhang, Y., Tuk, M. A., & Klesse, A.-K. (2024). Giving AI a human touch: Highlighting human input increases the perceived helpfulness of advice from AI coaches. *Journal of the Association for Consumer Research, 9*(3), 344–356. https://doi.org/10.1086/730710

Zhao, X., & Malle, B. F. (2022). Spontaneous perspective taking toward robots: The unique impact of humanlike appearance. *Cognition, 224*, 105076. https://doi.org/10.1016/j.cognition.2022.105076

Zhao, X., Phillips, E., & Malle, B. F. (2019). *How people infer a humanlike mind from a robot body*. https://psyarxiv.com/w6r24/download

Zheng, J. F., & Jarvenpaa, S. (2021). Thinking technology as human: Affordances, technology features, and egocentric biases in technology anthropomorphism. *Journal of the Association for Information Systems, 22*(5), 1429–1453. https://doi.org/10.17705/1jais.00698

Zilber, A. (2023). Ex-Google exec says AI-powered sex robots will seem 'alive' and eliminate need for 'another being'. *New York Post*. https://nypost.com/2023/07/18/ai-powered-sex-robots-will-seem-alive-and-negate-need-for-partner-ex-google-exec-says/

Zlatkov, D., Ens, J., & Pasquier, P. (2023). Searching for human bias against AI-composed music. In C. Johnson, N. Rodríguez-Fernández, & S. M. Rebelo (Eds.), *Artificial intelligence in music, sound, art and design. EvoMUSART 2023. Lecture notes in computer science* (Vol. 13988, pp. 308–323). Springer. https://doi.org/10.1007/978-3-031-29956-8_20

Index

D. B. Shank, *The Machine Penalty*, https://doi.org/10.1007/978-3-031-86061-4

The manufacturer's authorised representative in the EU is Springer Nature Customer Service Centre GmbH, Europaplatz 3, 69115 Heidelberg, Germany. If you have any concerns regarding our products, please contact ProductSafety@springernature.com

Printed and bound by CPI Group (UK) Ltd, Croydon, CR0 4YY
01/10/2026
02225912-0009